P9-EKJ-062

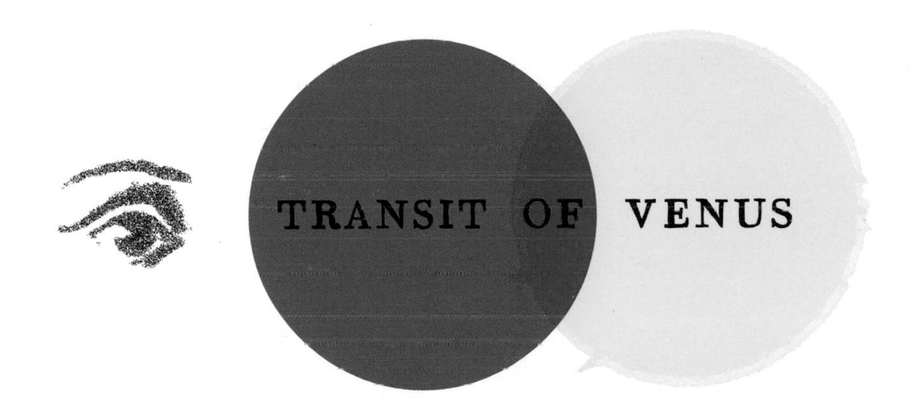
TRANSIT OF VENUS

The transit of Venus in 2004 captured in ultraviolet from the unique perspective of NASA's Sun-observing TRACE spacecraft. NASA/LMSAL

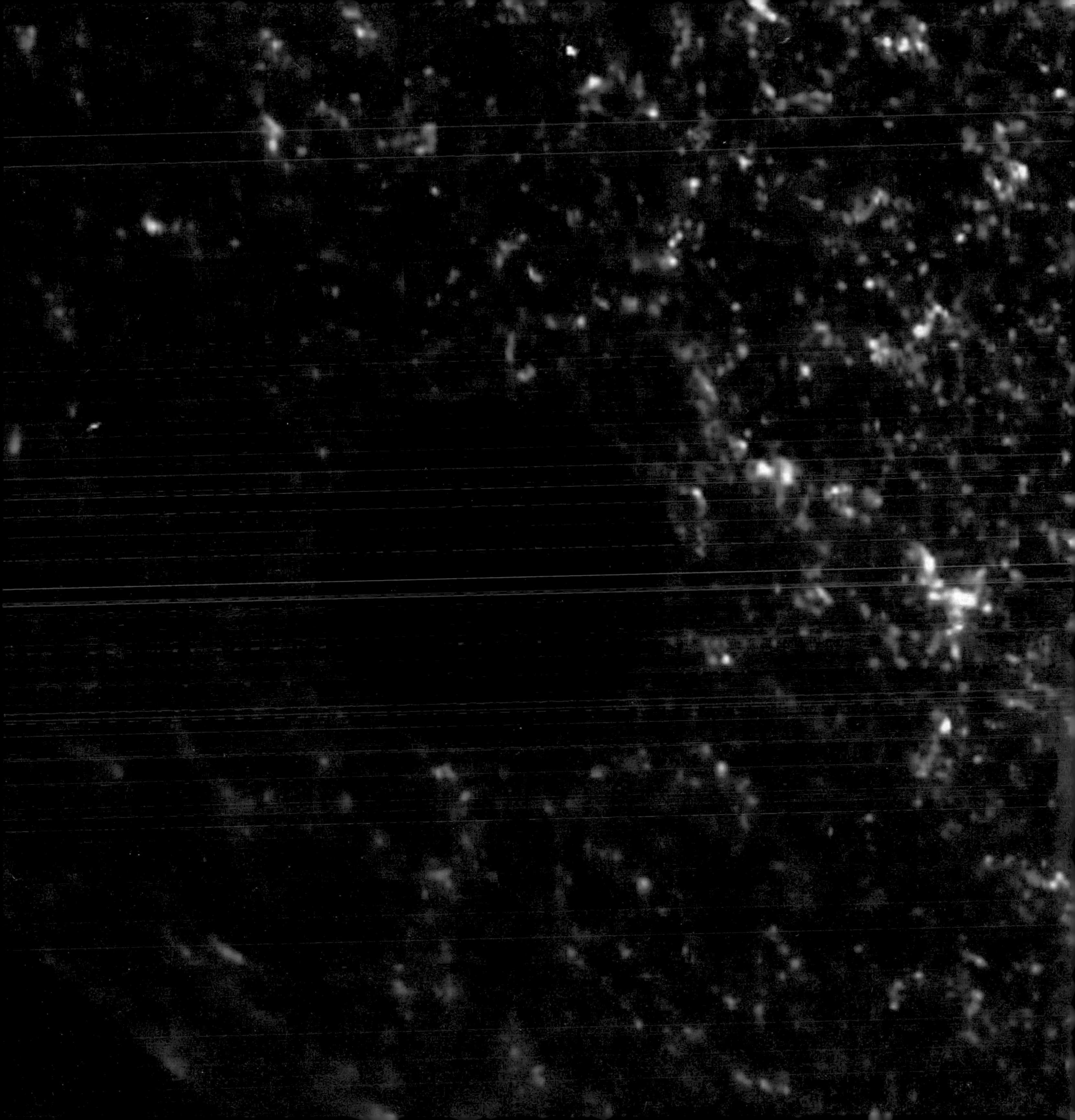

NICK LOMB

TRANSIT OF VENUS

1631 TO THE PRESENT

NEW YORK

TRANSIT OF VENUS: 1631 to the Present, by Nick Lomb

The Experiment
260 Fifth Avenue
New York, NY 10001-6408
www.theexperimentpublishing.com

Transit of Venus was first published in Australia by NewSouth, an imprint of UNSW Press. This first North American edition is published in association with the Powerhouse Museum, Sydney, Australia, and NewSouth Publishing.

This book is printed on paper using fibre supplied from plantation or sustainably managed forests.

The Experiment's books are available at special discounts when purchased in bulk for premiums and sales promotions as well as for fundraising or educational use. For details, contact us at info@theexperimentpublishing.com.

Library of Congress Control Number: 2011936834
ISBN 978-1-61519-055-3

Design by Di Quick
Manufactured in Canada

Distributed by Workman Publishing Company, Inc.
Distributed simultaneously in Canada by Thomas Allen and Son Ltd.

Published in North America in April 2012
10 9 8 7 6 5 4 3 2

Note on measurements
In this book measurements are generally given in metric and imperial units. For instances where no US equivalents are given, note that 1 km is about 0.6 miles, 1 metre is about 1.1 yards and 1 cm is about 0.4 inches. The value for the astronomical unit given in the 2011 Astronomical Almanac is equivalent to 92 955 807.24 miles.

CONTENTS

An artist's impression of the solar system showing the approximate relative sizes of the planets. Moving outwards from the Sun on the left, the planets are Mercury, Venus, Earth, Mars, Jupiter, Saturn, Uranus and Neptune, followed by the dwarf planet Pluto. Lunar and Planetary Institute NASA.

TRANSIT FAST FACTS

For Venus to transit across the Sun's face, the Earth and Venus must align in exactly the right way, making this a rare astronomical event (for more detailed information, see 'How a transit of Venus works', on pages 206–07).

- A TRANSIT OCCURS when a planet crosses the disc of the Sun as seen from the Earth.
- Only the two innermost planets, Mercury and Venus, can transit the Sun when seen from Earth.
- During a transit, Venus appears in silhouette as a dark circular spot about one-thirty-third of the width of the Sun.
- A transit of Venus can occur only in early June or early December.
- Eight years after a first transit, Venus and Earth are back in almost the same position and so there is the possibility of a second transit.
- If Venus crosses the southern part of the Sun at the first of a pair of transits, then it will cross the northern part at the following transit eight years later.
- If Venus crosses the northern part of the Sun at the first of a pair of transits, then it will cross the southern part at the following transit eight years later.
- If Venus crosses near the centre of the Sun, there will be no second transit eight years later. This situation will not occur until 3089.
- After the second of a pair of June transits, the next one occurs 105.5 years later, while after the second of a pair of December transits the next occurs 121.5 years later. For example, after the December transit of 1882 the next one was in June 2004.
- The pattern of transits repeats every 243 years.
- Since the invention of the telescope, transits of Venus have occurred in 1631 (unobserved), 1639 (the first observed), 1761, 1769, 1874, 1882 and 2004, with the next one in 2012.
- The following pair of transits of Venus will be on 11 December 2117 and 8 December 2125.

The paths of Venus and Earth around the Sun as seen from above.

INTRODUCTION

Every so often, the planet Venus does something remarkable. Its orbit brings it to a point directly between the Sun and the Earth, where it appears to us as a black dot moving across the bright disc of the Sun. This transit of Venus is rare, occurring in pairs eight years apart and then not for more than a hundred years; it has fascinated astronomers for centuries.

During the 18th and 19th centuries, astronomers and explorers set out on long and dangerous journeys to faraway places in order to observe this prized celestial event, often enduring great hardships along the way. The French astronomer Guillaume Le Gentil, for example, spent more than 11 years away from home trying, with varying degrees of success, to observe the two transits of 1761 and 1769 from southern Asia and the Indian Ocean. And, famously, the English explorer Lieutenant James Cook sailed to Tahiti for the transit of 3 June 1769, a voyage that led to his mapping the whole of New Zealand and the east coast of Australia and claiming these lands for the British Crown.

The attempts to observe those transits of past centuries were some of the earliest scientific expeditions. They led, for the first time, to international cooperation between scientists in planning and assisting observations, even in some cases while their countries of origin were at war.

Many people will recall the excitement of the transit of Venus on 8 June 2004, the first transit since 1882. Seeing Venus in front of the Sun for the first time in our lifetimes was a powerful experience, helping us to connect with the history and significance of this rare event, although

The cover of Henry Chamberlain Russell's *Observations of the Transit of Venus, 9 December, 1874.* Powerhouse Museum Research Library

Hemispheric view of Venus produced by the Magellan spacecraft. See box on the Magellan mission on pages 166–67. NASA.

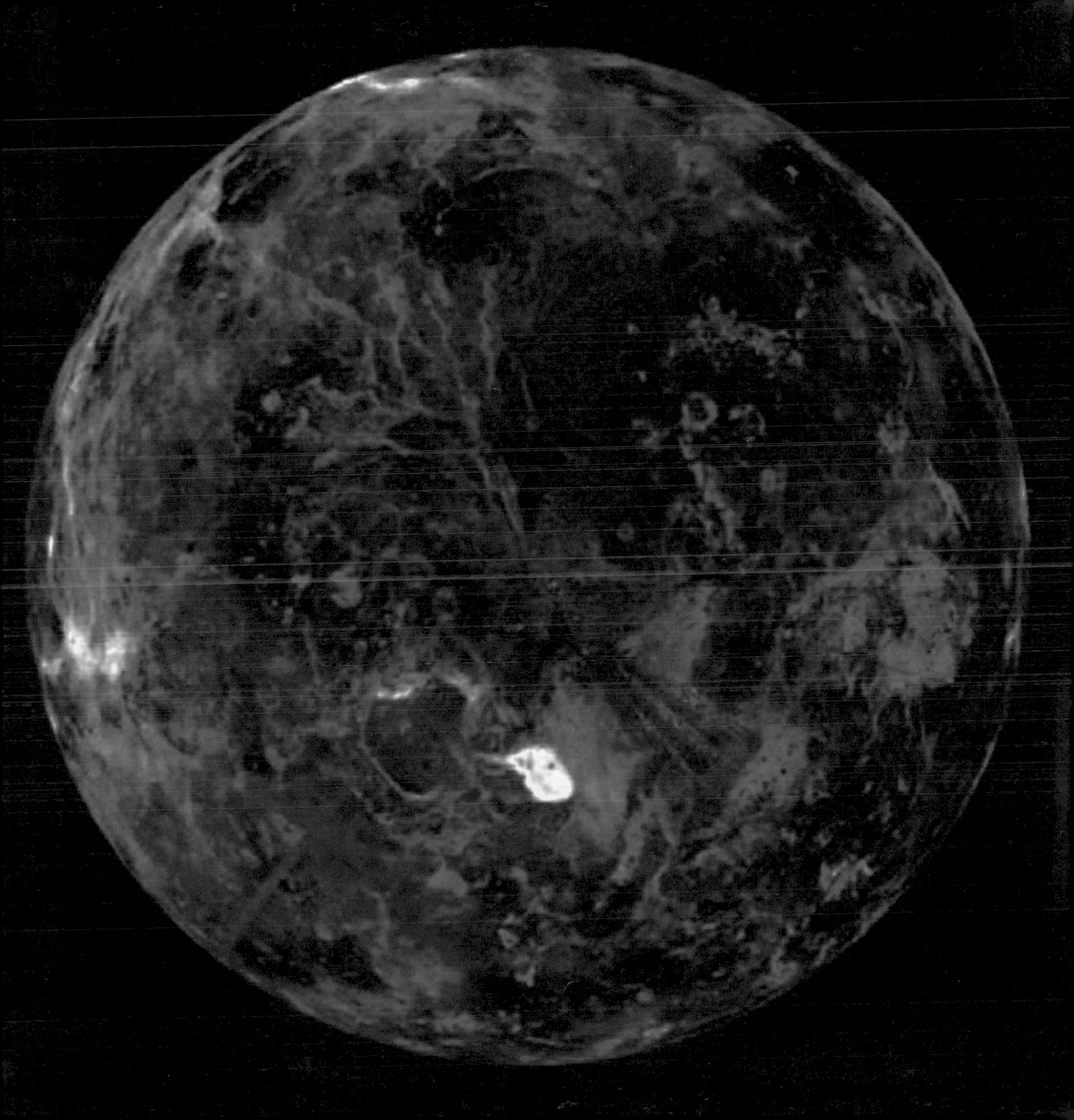

The Opera House and Sydney Harbour make a spectacular backdrop as four planets rise in the eastern pre-dawn sky. Venus is the brightest of the triangle, with Jupiter on the left and Mercury on the right. Mars is below, but almost lost in the twilight. Sydney Observatory, Sydney, Australia. May 2011. Photo: Melissa Hulbert

viewing the 2004 transit was far easier than it had been for the adventurous astronomers of previous centuries.

As I write these words, the second of the pair of 21st-century transits of Venus is fast approaching. This transit, the last until the year 2117, will take place on 5 or 6 June 2012 (depending on your location) and people on most continents will have the opportunity to see all or at least part of the event. The best observing locations will be in Australia, New Zealand, Japan, Korea and nearby countries. From there, observers will be able to see the transit from the time the planet moves in front of the disc of the Sun until it leaves it about six and a half hours later. From most of the continental United States, people will be able to see the Sun set with Venus still crossing its disc, while those in Europe wanting to see the event will need to rise early, as from there Venus will already be moving across the disc of the Sun at sunrise.

WHY ALL THE FUSS?

Why were astronomers of the 18th and 19th centuries so keen to observe the transit of Venus? The reason was that it provided a way of measuring the distance of the Earth from the Sun. That distance, known to astronomers as the astronomical unit (AU), is one

VENUS: A VISION OF HELL

ALTHOUGH VENUS IS NAMED after the Roman goddess of love, the conditions on its surface are anything but seductive. Our near neighbour is a hellishly hot, barren and inhospitable world.

A visiting astronaut could expect a ferocious welcome: he or she would be simultaneously fried alive by the 464°C temperatures (hot enough to melt lead) and crushed by the huge atmospheric pressure, which is 92 times that on Earth.

With a radius of 12104 kilometres, Venus is almost the same size as the Earth, but there the similarities end.

The planet's thick atmosphere is mainly made up of carbon dioxide with clouds of sulphuric acid that produce frequent electrical storms. This dense blanket of carbon dioxide prevents heat leaving the planet, which is why its surface is the hottest among all the planets in the solar system, even hotter than that of the innermost planet, Mercury.

Because the thick atmosphere has also prevented us from seeing the planet's surface with telescopes or satellites, NASA's Magellan spacecraft was sent to orbit the planet while scanning it with radar. Between 1990 and 1994, Magellan succeeded in mapping 98 percent of the surface, and showed that lava and volcanoes – hundreds of them – dominate the landscape (see box on pages 166–67).

An artist's impression of an electrical storm on Venus. NASA/ESA 2007-137

of the most fundamental quantities in astronomy. Without knowing that distance, astronomers could not make any headway in understanding the Sun because they could not calculate its true brightness, size or mass. The mass of the Sun is established by measuring its gravitational pull on Earth, which is related to the mass of the Sun divided by the square of its distance. So, if astronomers underestimated the distance by, say, a factor of two, the calculated mass would be four times less than the true value.

Astronomers knew that an accurate value of the distance from the Earth to the Sun would also allow them to work out the size of the solar system and the distances between each of the planets circling the Sun. In the early 1600s, the great German astronomer Johannes Kepler had determined the ratios of the distances of the planets from the Sun, so that, for example, it was known that the planet Jupiter was 5.2 times the distance of the Earth from the Sun. This meant that, if one distance could be reliably established, all distances in the solar system would be known.

Possibly the most important reason why the Sun's distance was crucial to astronomers is that the Earth's yearly path can provide a baseline for measuring the distances of stars other than the Sun. By observing the tiny shift in position of a star at two ends of the Earth's path, say in winter and summer, astronomers can determine its distance from us. The first star to have its distance established in this way was 61 Cygni in 1838, while the next in 1839 was that favourite star of the southern sky, Alpha Centauri, one of the two pointers to the Southern Cross. For this technique to work, the fundamental distance of the Earth to the Sun needed to be known.

Even today, the distances of nearby stars are determined using the same technique of measuring the tiny shifts in their position in the sky during the course of a year, although the methods modern astronomers use are much more accurate than those used in the past. In 2012, the European Space Agency plans to launch the Gaia spacecraft with the aim of measuring the distances and other parameters of a phenomenal one billion stars during its five-year or longer mission. The results from the spacecraft will greatly boost our knowledge of stars in our galaxy, the Milky Way.

Modern astronomers have developed a complex distance scale that takes us from the nearest star, Proxima Centauri, part of the Alpha Centauri system (so close that light from it takes only four and one third years to reach us) to the most distant galaxy known (so far away that its light has taken

13.2 thousand million years to arrive). This distance scale involves a number of different techniques, but each fundamentally depends on the determinations of the distance of the nearest stars and of the Earth to the Sun.

For all these reasons, the determination of the exact value of the distance to the Sun was one of the most important tasks facing astronomers of the 18th and 19th centuries. Indeed, Sir George Airy, who was Astronomer Royal at Greenwich Observatory, London, for much of the 19th century, called the determination of that value, 'the noblest problem in astronomy'.

Transits of Venus offered a way of tackling the problem. During a transit, Venus can come as close as 38 million kilometres to Earth, closer than any other planet. What's more, at those times the planet is clearly visible in silhouette, with the Sun's bright surface providing a sort of ruler or protractor to allow measurement of the very small angles involved in the distance determinations.

Astronomers from many places, supported by their countries mainly for reasons of national prestige, eagerly observed the transits of 1761, 1769 and 1874 and, with somewhat diminished enthusiasm, the transit of 1882. This book recounts some of their adventures, successes and disappointments. As we will see, the results of their huge efforts pinned down the value of the Sun's distance to within a few million kilometres of the present-day accepted value.

How did modern astronomers reach that accepted value? They still use the planet Venus, but now they do it by sending radar signals towards the planet and waiting for the echo to arrive back. When Venus is at its closest to Earth this wait is about five minutes. Once the distance to Venus is determined, the astronomical unit is easily found using the known ratio between the distances of Earth and Venus from the Sun. Separate groups of astronomers located at the 76-metre radio telescope at Jodrell Bank near Manchester in the UK, at NASA's 70-metre receiver at Goldstone in New Mexico in the United States, and in the Soviet Union first used this technique in 1961.

The current value for the astronomical unit is so accurate that its exact value depends on the precise definition of a second that is used, as this affects the value of the speed of light. The 2011 *Astronomical Almanac*, an annual publication that is the fundamental reference for the motion of the Sun, the planets and the values of astronomical constants, lists its value as precisely 149597870700 metres with a possible error of only 3 metres.

A crescent Venus shining in daytime blue sky as seen through the circular field of view of a 25-centimetre Meade telescope. The photograph was taken using a small digital camera held to the telescope eyepiece with a shutter speed of 1/250 seconds and an ISO speed of 80. Photo: Nick Lomb

An artist's impression of the Gaia spacecraft that is to be launched in 2012 with the aim of measuring the distances to a billion stars. ESA

One of the best-known comments about the transit of Venus was made in August 1882 by US Naval Observatory astronomer William Harkness, who observed the 1874 transit from Hobart, Australia. As we approach the 2012 transit of Venus, we too might like to think about what the world will be like for our descendants when the next transit takes place in the northern winter and southern summer of 2117:

> Transits of Venus usually occur in pairs; the two transits of a pair being separated by only eight years, but between the nearest transits of consecutive pairs more than a century elapses. We are now on the eve of the second transit of a pair, after which there will be no other till the twenty-first century of our era has dawned upon the earth, and the June flowers are blooming in 2004. When the last transit season occurred the intellectual world was awakening from the slumber of ages, and that wondrous scientific activity which has led to our present advanced knowledge was just beginning. What will be the state of science when the next transit season arrives God only knows. Not even our children's children will live to take part in the astronomy of that day. As for ourselves, we have to do with the present ...

A SPOT OF UNUSUAL MAGNITUDE

1639

One Sunday in late 1639, in the little village of Much Hoole near Liverpool, England, Jeremiah Horrocks set about preparing for a great event, the first transit of Venus across the face of the Sun ever to be observed by human beings. Horrocks arranged a telescope so that it projected the Sun's image onto a piece of paper in a darkened room. The image was only 6 inches (150 millimetres) wide — the size of his room did not allow anything larger — and on the piece of paper he divided the circumference into degrees and the diameter into 30 equal parts. Now the young man waited for events to take their course.

In the church at Much Hoole, a stained-glass window commemorates Jeremiah Horrocks. Photo: Val and Andrew White.

ORTVS
AVRORA

< By 1601, when the great German astronomer Johannes Kepler moved to Prague to join another great astronomer, Tycho Brahe, the town's famous astronomical clock had been operating for almost two centuries. As well as the time, the clock shows much astronomical detail, including the position of the Sun and the Moon. Photo: Nick Lomb

A statue of astronomers Johannes Kepler and Tycho Brahe in Prague. Photo: Nick Lomb

The first half of the 17th century in Europe was a time of political, commercial, cultural and scientific ferment. Europeans were moving further afield, with the establishment of new settlements in the New World. In November 1620, the *Mayflower* arrived in North America carrying some 100 pilgrims seeking religious freedom.

Although long sea voyages were risky, there were many brave enough to sail for reasons of trade or exploration. In 1606, a small Dutch ship, the *Duyfken*, under the command of Willem Janszoon, made the first recorded European landing on the continent we now know as Australia.

Some of William Shakespeare's best known plays were first performed in this period: *Twelfth Night* in 1601, *Othello* in 1602, *King Lear* in 1605 and *Macbeth* in 1606. Shakespeare himself died in 1616, leaving his 'second best bed' to his wife Anne. The Spanish novel *Don Quixote* appeared in 1605 while painters such as Peter Paul Rubens and Frans Hals were using oil paint for the first time.

In the decades leading up to the 1639 transit of Venus across the Sun, most European powers were embroiled in the Thirty Years' War, which dragged on from 1618 to 1648. Of complex, mainly religious origins, the conflict led to famine, disease and death on a large scale, as well as financial ruin for some of the combatant nations.

These dangerous times saw some exciting developments in science, particularly in astronomy. In 1609, the Italian scientist Galileo Galilei first pointed his tiny telescope to the sky and made some amazing discoveries, including four moons circling the planet Jupiter. This was an astonishing observation, because it showed that the Earth was not necessarily the centre of all motion in the Universe as had been believed up to that point. Such unorthodox ideas led to Galileo's 1633 trial and conviction for heresy by the Catholic Church.

KEPLER PAVES THE WAY

At around the same time, the great German mathematician and astronomer Johannes Kepler (1571–1630) was preparing the ground for the astronomers who were to follow, including Jeremiah Horrocks. In 1601, Kepler became Imperial Mathematician to Emperor Rudolf II in Prague, the capital of Bohemia. Rudolf had broad interests and he drew a variety of scholars and artists to the city. These included astrologers and alchemists as there was no clear distinction between the occult and science at the time. Kepler himself straddled the two worlds – one of his duties was to cast horoscopes for the Emperor.

Kepler had come to Prague to work with the famous observational astronomer Tycho Brahe on the interpretation of Brahe's observations. Together the two astronomers undertook to calculate a new set of tables on the positions and motions of the stars and planets, to be called the *Rudolphine Tables* after the Emperor. When Brahe died only a few months after Kepler's arrival, in October 1601, Kepler succeeded him as Imperial Mathematician.

Part of Kepler's work was to interpret Brahe's observations of Mars. After numerous failed attempts to understand the shape of the planet's orbit, Kepler tried matching Brahe's observations of it with an ellipse – an oval shape with two focal points – and found a perfect match. He generalised this simple and clear result to become his first law of planetary motion: all planets move in ellipses with the Sun at one focus point. His second law was one that he had deduced earlier: planets sweep out equal areas in equal times. This second law means that a planet moves faster when near the Sun than away from it. In 1609 Kepler published these two laws in *Astronomia Nova (A New Astronomy)*.

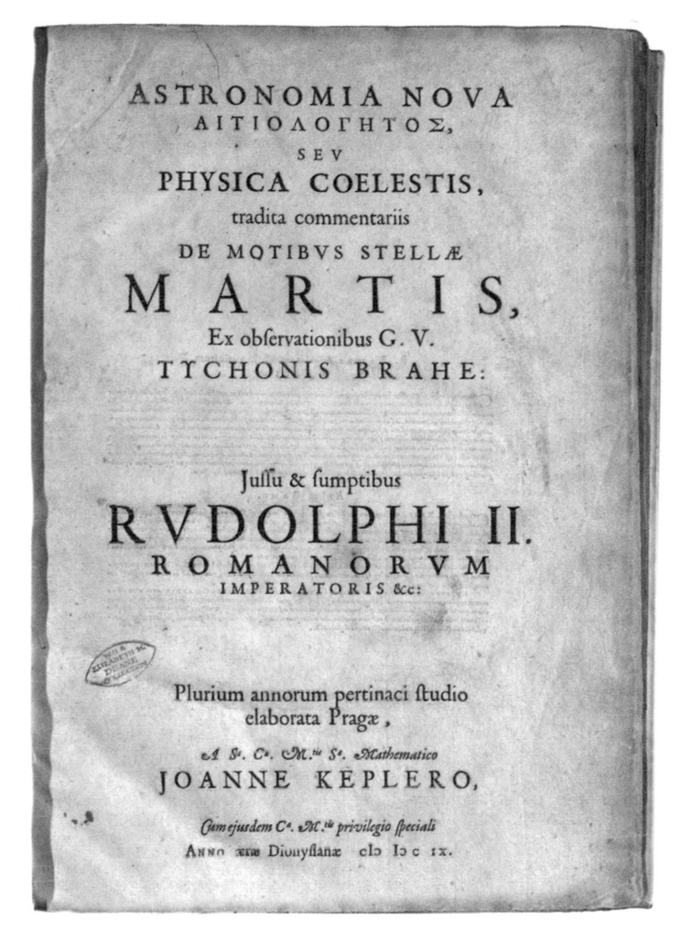

ASTRONOMIA NOVA
ΑΙΤΙΟΛΟΓΗΤΟΣ,
SEV
PHYSICA COELESTIS,
tradita commentariis
DE MOTIBVS STELLÆ
MARTIS,
Ex obſervationibus G. V.
TYCHONIS BRAHE:

Juſſu & ſumptibus
RVDOLPHI II.
ROMANORVM
IMPERATORIS &c:

Plurium annorum pertinaci ſtudio
elaborata Pragæ,
A S. C. M.tis S. Mathematico
JOANNE KEPLERO,
Cum ejusdem C. M.tis privilegio ſpeciali
ANNO æræ Dionyſianæ clɔ Iɔ c ix.

The title page of Johannes Kepler's *Astronomia Nova (A New Astronomy),* published in Prague in 1609. The University of Sydney Rare Books and Special Collections Library

Kepler was not fully satisfied with these laws and tried to find some kind of pattern that would link the orbits of different planets. After many years of fruitless work, the relationship that became his third and most important law came to him on 18 May 1618, by which time he was living and working in the Austrian town of Linz: the square of the time a planet takes to go around the Sun is proportional to the cube of its distance from the Sun. This may seem a complicated and esoteric relationship, but it allows astronomers to use their observations of the times the different planets take to circle the Sun to calculate their relative distances from the Sun. Kepler's third law meant that, once the actual distance of one planet was known, the distances of the other planets could also be deduced. As we will see later, the great importance of the transits of Venus to astronomers of the 18th and 19th centuries was that it could provide that one key distance needed to establish the distances of all the planets.

Kepler completed the the *Rudolphine Tables* in 1623 and it was published four years later. In completing the huge number of calculations needed for the tables, Kepler was helped by the newly devised calculation method of logarithms, a method that would continue to be widely used into the late 1960s, when it was finally made redundant by the widespread use of electronic calculators.

In 1604, Kepler saw a bright new star in the sky that we now know was a supernova or exploding star. The background image shows the still expanding remnant of this exploding star in a variety of wavelengths from X-rays to infrared. NASA, ESA, R. Sankrit and W. Blair (Johns Hopkins University)

The horoscope cast by Kepler for General Albrecht von Wallenstein in 1608. Getty Images

On the cusp of the modern era, Johannes Kepler made brilliant contributions to science, yet still cast horoscopes.

JOHANNES KEPLER

KEPLER WAS A BRILLIANT mathematician and scientist, who lived at a time when mystical and supernatural beliefs were still very much current. He made great advances in our understanding of the solar system and the motions of the planets, yet he also cast horoscopes. He believed that observations and mathematics would explain the workings of the Universe, yet his results often led him to ask questions that seem ridiculous in the modern era.

Kepler was born on 27 December 1571 in the town of Weil der Stadt near Stuttgart in what is now Germany. His introduction to astronomy came at the University of Tübingen. Initially he learned about the officially sanctioned conception of the motions of the planets as proposed by the Greek astronomer, Claudius Ptolemy, in the 2nd century AD. According to the Ptolemaic model, all the planets including the Sun and the Moon orbited the Earth while moving in a combination of circular motions. When Kepler later heard about the revolutionary ideas of Polish astronomer Nicolaus Copernicus, who proposed that the planets circle the Sun and not the Earth, he was immediately convinced.

Kepler then turned his mind to the relative spacing of the planets in the Copernican system. Why were they arranged with increasing distances between them and why with those particular values? To modern scientists, this arrangement is a random result of the history of the solar system, but to Kepler it became an obsession. He attempted to solve the problem with a geometric construction involving regular solids inserted into spherical shells and obtained a close enough match to the actual spacing to be encouraged.

In 1601, he moved to Prague to work with the astronomer Tycho Brahe, who was well known for his precise observations of the heavens. Here, Kepler determined the path of the planet Mars around the Sun and established his first two laws of planetary motion. In 1612, he left to become professor of mathematics at Linz in Austria.

Kepler's personal life became troubled at this time, when one of his three children died of smallpox, shortly followed by the death of his wife. Kepler considered 11 possible new wives before settling on number five on his list, Susannah, in 1613. Seven years later, Kepler's mother was charged with witchcraft and condemned to be tortured, a fate she only avoided thanks to the intervention of her son, although she died soon after her release from prison.

After the publication of his important work, the *Rudolphine Tables* in 1627, Kepler was employed by the successful general Albrecht von Wallenstein. Kepler had already cast a horoscope for Wallenstein back in 1608 and he was expected once again to provide astrological information and advice, a task he performed reluctantly. Although he also retained his post as Imperial Mathematician to the Emperor in Prague, Kepler had great difficulty getting paid. He made his final journey, to the German town of Ratisbon or Regensburg in an attempt to get the money he was owed. There he contracted a fever that led to his death on 16 November 1630 (Gregorian calendar). He was buried in Regensburg, but his tombstone was destroyed when the town was occupied by Swedish soldiers three years later.

JOANNIS KEPPLERI
MATH. CÆS.
ADMONITIO
AD ASTRONOMOS,
RERUMQVE COELESTIUM
STUDIOSOS,
De raris mirisq; Anni 1631.
Phænomenis,
VENERIS PUTA ET MERCURII
in Solem incursu:
Excerpta
Ex EPHEMERIDE Anni 1631. & certo AUTHORIS consilio huic præmissa, iterumq; edita
à
JACOBO BARTSCHIO
Lauba Lusato, Phil. & Med. D.

FRANCOFURTI
Apud Godefridum Tampachium.
ANNO M.DC.XXX.

While calculating astronomical events for the years 1629–30 based on the completed *Rudolphine Tables*, Kepler made the extraordinary discovery that in 1631 both Mercury and Venus were likely to cross or transit in front of the Sun. Kepler was particularly attuned to transits of Mercury as he had claimed to have seen one – mistakenly as it turned out – in 1607. Once sunspots were discovered with the newly invented telescope two years later, he realised and admitted his error.

For the predicted transits of 1631, Kepler wrote an 'admonition' to astronomers to observe the events, including instructions on how and when to observe. After his death in 1630 his student and son-in-law Jacob Bartsch published the admonition as a separate pamphlet. Though the Mercury transit was to take place on 7 November 1631, Kepler suggested that astronomers begin their observations a day earlier and finish a day later since he was not sufficiently confident of the path of the planet. The transit of Venus was calculated to occur on 6 December 1631.

Johannes Kepler's *Admonition* published by Jacob Bartsch at Frankfurt, Germany, in 1630. Herzog August Bibliothek Wolfenbüttel

The frontispiece to the *Rudolphine Tables* prepared by Johannes Kepler. It pays homage to four great astronomers of the past: the Greek astronomers Hipparchus and Ptolemy, plus Nicolaus Copernicus and Tycho Brahe. The University of Sydney Rare Books and Special Collections Library

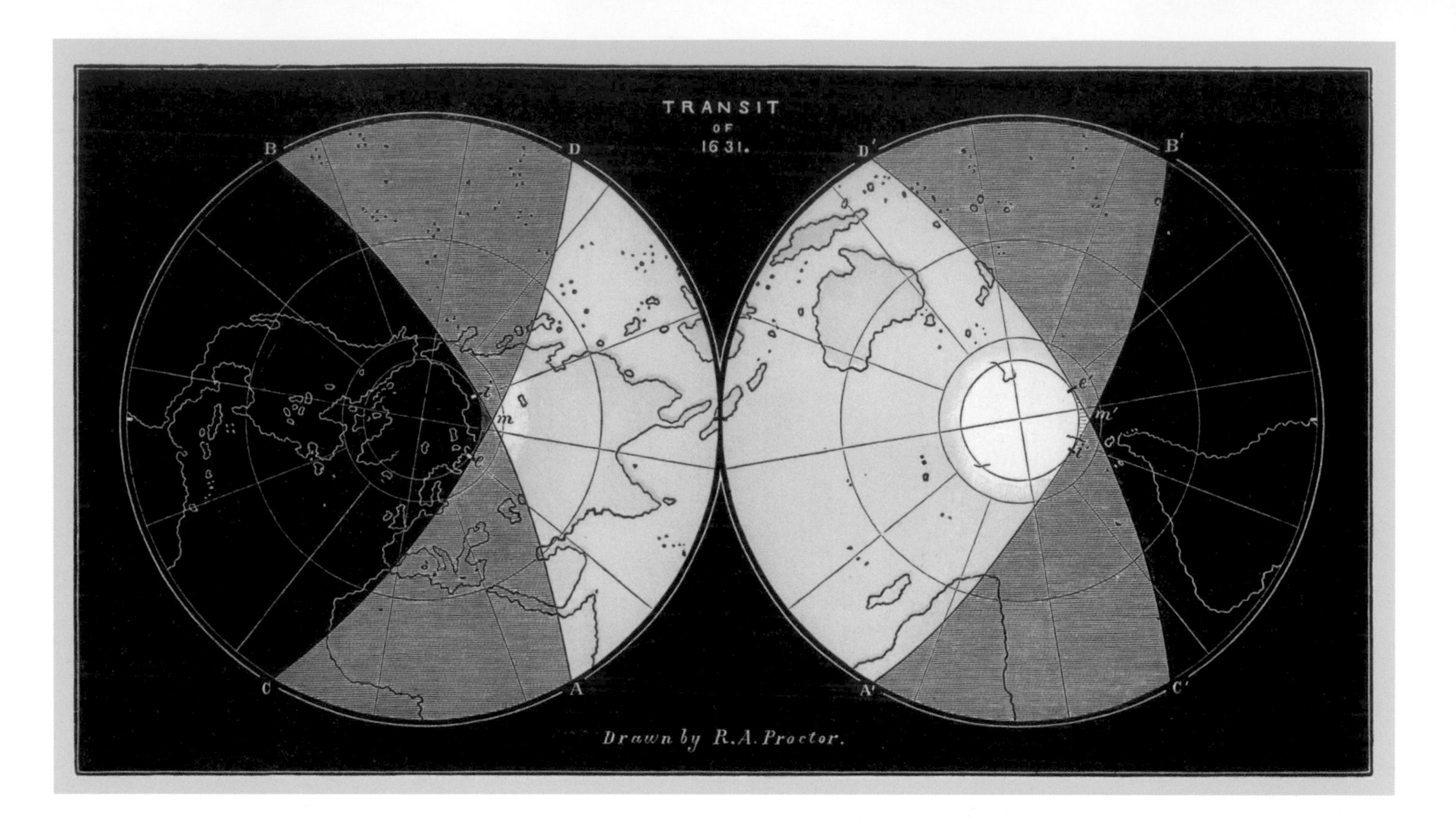

The visibility of the 1631 transit of Venus, with the northern hemisphere on the left and the southern hemisphere on the right. The brightest area denotes the regions from which the transit was fully visible, the shaded areas indicate regions from which the beginning or end of the transit was visible, and the darkest areas are the regions from which the transit could not be seen. Map from *Transits of Venus* by RA Proctor, 1874. Brian Greig collection

GASSENDI AND HORROCKS

A number of astronomers in Europe took up the challenge of trying to observe the transits. Only a few succeeded in observing the Mercury transit and of these only the observations of Pierre Gassendi in Paris were published. No one is known to have seen the Venus transit, which occurred at night-time in Europe.

Pierre Gassendi (1592–1655) was a French priest who was also a philosopher and a scientist with an interest in astronomy. It was Gassendi who gave the name Aurora Borealis to the Northern Lights that are visible in far northern latitudes – 'aurora' refers to the Roman goddess of Dawn, while 'borealis' comes from the Greek name for the northern wind.

Gassendi was used to observing sunspots so it was not difficult for him to prepare to observe the transit of Mercury. Using a telescope, he projected an image of the Sun onto a piece of paper in a darkened room. On the paper, he drew a circle matching the size of the image, which was eight Parisian inches in width, equivalent to about 22 centimetres. To estimate the angular size of Mercury should he see it, he divided the width of the circle into 60 equal parts so that each division was equal to 30 seconds of arc. Finally, to record the time of each observation, he arranged for an assistant to be in the room below to measure the altitude of the Sun with a quadrant each time he heard the thump of Gassendi stamping his foot on the floor.

Gassendi described his observations in detail in his *Mercurius in sole visus et Venus invisa*, published in 1632. He tells us that he planned to begin his wait for Mercury two days early to maximise his chance of seeing the event, but rain prevented his seeing the Sun all day on 5 November. The next day was not much better with fog only allowing a few glimpses of the Sun. On 7 November, the day for which Kepler had predicted the transit, Gassendi managed to see the Sun on a number of occasions starting from just before 9 am. On his first view that day he saw a small dark spot that was far smaller than he had expected Mercury to be. The width of the spot, he said, 'hardly exceeded half of the divisions marked' and he assumed it was a sunspot.

Gassendi decided to use the sunspot to measure the motion of Mercury should it appear. But the supposed sunspot kept moving across the Sun far more quickly than was possible for such a spot and finally Gassendi realised that he was actually looking at the sought-after Mercury. He then used the scale he had prepared on the paper projection screen to establish that the width of the planet's silhouette was 20 seconds of arc. He made the pre-arranged thump on the floor, but unfortunately the assistant had left the room below convinced that nothing would be seen that day. Eventually, the assistant returned and Mercury's departure, or egress, from the disc of the Sun was timed at 10.28 am.

This first ever observation of a planet transiting the Sun was hugely important. Gassendi was ecstatic, writing to a colleague in Tübingen in Germany, 'I have found him, I have seen him where no one has ever seen him before!' The observation allowed astronomers to refine Kepler's calculations for the path of Mercury around the Sun. Even more importantly,

Horrocks saw Venus pass across the face of the Sun, ... 'a spot of unusual magnitude and of a perfectly circular shape'.

THE FIRST PERSON TO HAVE observed and left an account of a transit of Venus is a mysterious figure. Jeremiah Horrocks' achievements were immense, but there are many unanswered questions about his short life (he died aged only 22). He was probably born in Toxteth, now a suburb of the English city of Liverpool, in 1619. His father may have been a watchmaker. We do know he was educated in Cambridge as he is recorded as entering Emmanuel College there as a 'sizar' on 18 May 1632.

A sizar was a poor student who was given free tuition in return for carrying out menial servant-like duties. Horrocks appears to have been just 13 years old when he entered the college, but this was not unusual at the time. It was at Cambridge that he became interested in astronomy. This interest, as well as the considerable mathematical skill that he acquired, is surprising as it is unlikely to have been part of his official university studies. Soon after leaving Cambridge, he met William Crabtree, a Manchester cloth merchant, who was to become his closest collaborator. Although Johannes Kepler had predicted the 1631 transit, he had failed to realise that transits came in pairs. It was Horrocks who realised, only a month before the event, that a second transit was likely in 1639.

He probably observed the 1639 transit from the village of Much Hoole, to the north of Liverpool, where he lived for a while after leaving Cambridge. In his description, he says only that he made the observation, 'in an obscure village where I have long been in the habit of observing, about fifteen miles to the north of Liverpool'. There he saw the planet pass across the face of the Sun, 'a most agreeable spectacle ... a spot of unusual magnitude and of a perfectly circular shape'.

We do not know how Horrocks came to be living in Hoole, although that has not prevented people from speculating. Many writers have given him the title 'Reverend' and assumed he was the local curate. The assumption probably originated from a line in his transit observations, where he wrote that at one point he was 'called away in the intervals by business of the highest importance which, for these ornamental pursuits, I could not with propriety neglect'. As it was a Sunday, this could be interpreted to mean he was a clergyman. However, historian Allan Chapman points out that the then 20-year-old Horrocks was four years too young to have been ordained. Chapman suggests Horrocks was likely to have been a schoolmaster carrying out the normal Sunday observance of a religious person in the 17th century.

Over five years, Horrocks carried out an intensive campaign of astronomical observations, building and designing his own instruments as these were not yet being made. One of his instruments, which he called an Astronomical Radius, consisted of two graduated wooden rods with which he could measure the angle between two stars or, with a special version, the angular width of the Sun's disc.

In January 1641, Horrocks had arranged to visit his friend and collaborator William Crabtree, but sadly he died on the 4th of the month, the day before the planned meeting. Nothing is known of the circumstances of his death.

Portrait of Jeremiah Horrocks painted circa 1903 by JW Lavender, in the collection of the Astley Hall and Art Gallery, Chorley. Wikimedia Commons

Horrocks and his observations have not been forgotten, and he is especially remembered when transits of Venus occur. In 1874, a fund was set up to erect a tablet in London's Westminster Abbey, a rare honour. The list of subscribers – headed by the eminent astronomers John Couch Adams, Professor of Astronomy at Cambridge, and Sir George Airy, the Astronomer Royal – reads like a who's who of British astronomy of the late 19th century.

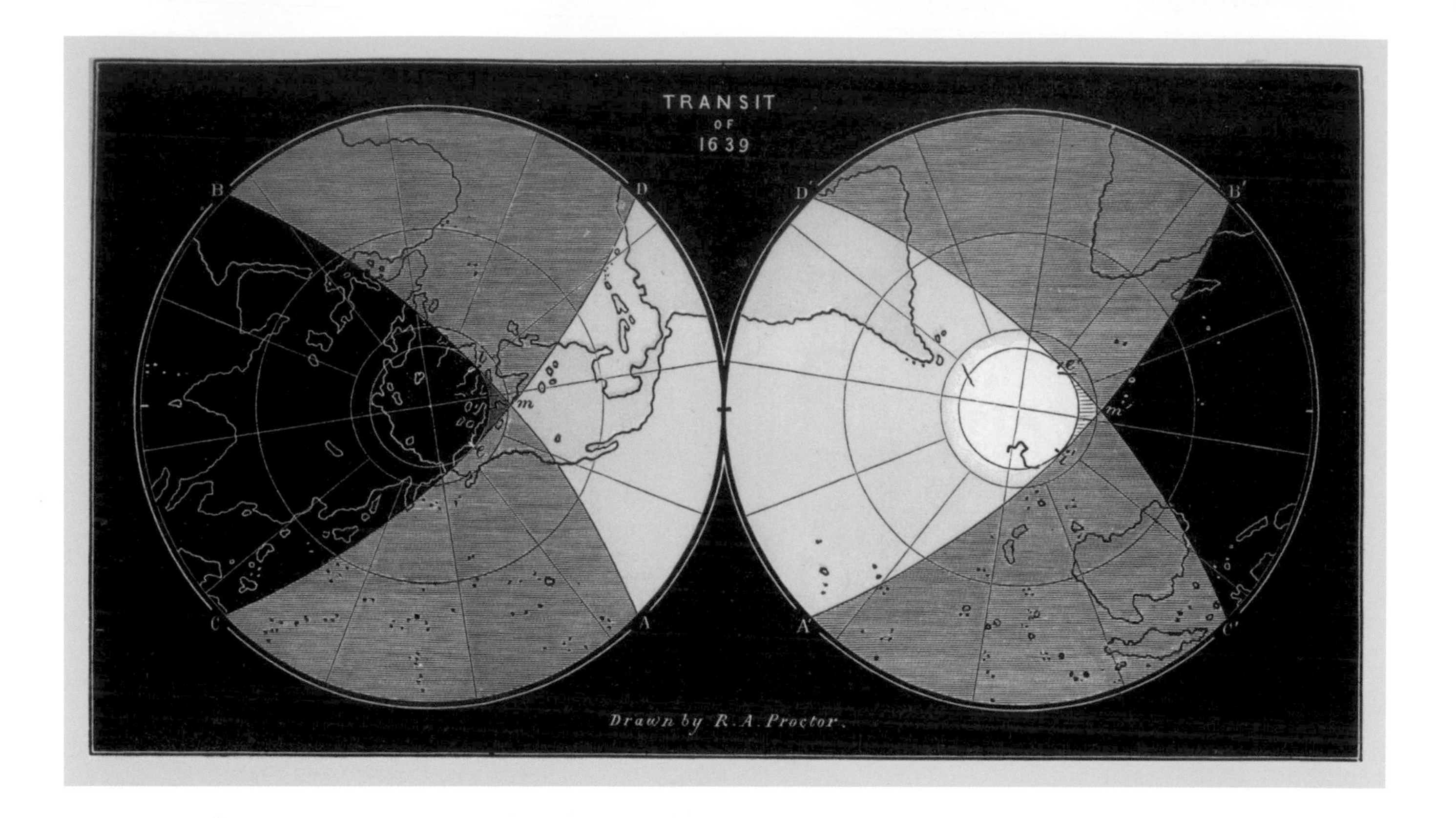

the small value for Mercury's angular size as determined by Gassendi implied a smaller angular size for the other planets than previously estimated and hence a larger solar system.

Although Johannes Kepler had correctly predicted the observed transit of Mercury (and the unseen transit of Venus) in late 1631, he had incorrectly stated that the next transit of Venus would not take place for over a century, not until 1761. Somehow he had missed the possibility of a second transit eight years later. It was the young English astronomer Jeremiah Horrocks who fortuitously realised only a month before the event that a transit was likely in 1639.

The visibility of the 1639 transit of Venus, with the northern hemisphere on the left and the southern hemisphere on the right. The brightest area denotes the regions from which the transit was fully visible, the shaded areas indicate regions from which the beginning or end of the transit was visible, and the darkest areas are the regions from which the transit could not be seen. Map from *Transits of Venus* by RA Proctor, 1874. Brian Greig collection.

On 24 November in the Julian calendar still in use in England until 1752, or on 6 December 1639 in the Gregorian calendar, Venus would be in the same direction as the Sun. Horrocks consulted all the astronomical tables he could find, including Kepler's, to try to find out when Venus would cross the plane in which the Earth circles the Sun. Some of the tables, combined with Horrocks' own observations, suggested the crossing would also happen on the 24th. When Horrocks realised another transit of Venus was possible, he alerted his friend William Crabtree – who would observe the transit separately – and made careful preparations for observing the event, using a set-up similar to Gassendi's.

Horrocks wrote up his successful observations of the transit together with his deductions and conclusions from them in a manuscript titled *Venus in sole visa*, possibly influenced by the title of Gassendi's work on the Mercury transit. The manuscript was eventually published in 1662, long after Horrocks' death in 1641. Horrocks' account tells us that he began observing the Sun the day before the predicted transit, just in case there was an error in the calculations. On the day itself, the sky was fortunately clear and he watched from 9 am to 1 pm with only a few breaks. After 'business of the highest importance' kept him away until 3.15 pm, he says:

> I then beheld a most agreeable spectacle, the object of my sanguine wishes, a spot of unusual magnitude and of a perfectly circular shape, which had already fully entered upon the Sun's disk on the left, so that the limbs of the Sun and Venus were precisely coincided, forming an angle of contact. Not doubting that this was really the shadow of the planet, I immediately applied myself sedulously to observe it.

The Sun set only half an hour later, but that short space of time greatly increased both our knowledge of the solar system and Horrocks' reputation. His observations allowed him to calculate improved values for the parameters that define the path of Venus around the Sun. The sharp blackness of Venus against the brightness of the Sun allowed him to clearly demonstrate that the planets were solid objects shining by reflected light and not, as some people of the time believed, made of some unknown luminescent substance.

Most importantly, he noted that the angular size of Venus he had measured during the transit was smaller than other astronomers had previously assumed. Putting his measurement for Venus together with Gassendi's for Mercury, Horrocks estimated the distance of the Earth from the Sun at about 191 million kilometres. Though this value is too high and Horrocks' method of estimation

The front page of Edmond Halley's *Catalogus Stellarum Australium (Catalogue of the Southern Hemisphere Stars)*, published in 1679. Collection: Powerhouse Museum, Sydney

CATALOGUS
STELLARUM AUSTRALIUM
SIVE
Supplementum
CATALOGI TYCHONICI
EXHIBENS
Longitudines & Latitudines Stellarum
fixarum, quæ, prope Polum Antarcticum ſitæ, in Horizonte *Uraniburgico Tychoni* inconſpicuæ fuere, accurato calculo ex diſtantiis ſupputatas, & ad Annum 1677 completum correctas; cum ipſis Obſervationibus in Inſula S. *Helenæ* (cujus Latit. 15 *gr.* 55 *m.* Auſtr. & Long. 7 *gr.* 00 *m.* ad occaſum a *Londino*) ſumma cura & ſextante ſatis magno de Cœlo depromptis.

Opus ab Aſtronomis hactenus deſideratum.

Accedit Appendicula de rebus quibuſdam Aſtronomicis, notatu non indignis.

Authore *EDMUNDO HALLEIO*,
E Col. Reg. *Oxon.*

LONDINI:
Typis *Thomæ James* Typographi Mathematici Regii, Impenſis *R. Harford* ad Inſigne Angeli, in Vico vulgo dicto *Cornhil*, prope Mercatorium Londinenſe. Anno Chriſti
MDCLXXIX.

contains some assumptions we now regard as strange, it moved prior estimates greatly in the right direction.

HALLEY'S TRANSIT OF MERCURY

In 1677, another young English astronomer observed the transit of a different planet – Mercury – from an island in the Atlantic. Decades later, his reputation as a man of science established, Edmond Halley wrote of his experiences in impressive style:

> About forty years ago, whilst I was in the island of St. Helena, observing the stars about the south pole, I had an opportunity of observing, with the greatest diligence, Mercury passing over the disc of the Sun; and (which succeeded better than I could have hoped for) I observed with the greatest degree of accuracy, by means of a telescope 24 feet long, the very moment when Mercury entering upon the Sun seemed to touch its limb within, and also the moment when going off it struck the limb of the Sun's disc, forming the angle of interior contact; whence I found the interval of time, during which Mercury then appeared within the Sun's disc, even without an error of one second of time.

As he had managed to time the duration of the Mercury transit exactly, Halley suggested similar measurements should be possible during transits of Venus. Using those measurements, astronomers could measure the distance of the Sun, and therefore gauge the size of the solar system. This carefully thought-out idea, complete with practical detail, was published in *Proceedings of the Royal Society* in 1716. It was to capture the imagination of astronomers.

Although others, including the Scottish mathematician James Gregory, had previously hinted at the possibility of using the transits for this purpose, unlike Halley they had not provided a practical method for determining the Sun's distance.

In his famous paper, Halley explained his idea that the duration of the transit of Venus be measured from 'different parts of the earth'. Using a comparison of the duration from different places with even approximately known positions, it becomes a matter of simple geometry to determine the Sun's distance. Mercury, Halley said, was not suitable for the task as it was too far from the Earth and too close to the Sun, meaning there would be insufficient time differences between its transit durations as observed from different locations.

MERCURII TRANSITUS
Sub Solis disco, *Octob.* 28. Anno 1677, cum tentamine pro Solis Parallaxi.

Arum istud, & a mortalibus non nisi ter, (quod mihi scire contigit,) hactenus observatum Phænomenon transitus Mercurii sub Solis disco, mihi, in Insula Sanctæ *Helenæ* commoranti, felicius observare, quam cuivis alio Astronomo, contigit: *Gassendus* enim in transitu Anni 1631, & in hoc nostro Clarissimus *Gallet*, exitum solum spectaverunt; ingressu, huic sub densa nubium compagine, illi sub terra Orientali, latente; Atque imperfectius adhuc Anno 1661, inclytus ille *Hevelius Gedani*, & nostrates *Londini*, qui solo situ intra faciem Solarem sumpto contenti erant: Mihi primo & ingressus & egressus momenta accuratissime conspecta sunt, idque peculiari & insolito Cœli favore; erat enim nocte præcedente *Octobris* 28vum Cœli facies tristissima, cum vento valido, interdumque descendentibus Nubibus densa Nebula Insulæ summitates obvelavit; luce reversa, vento licet paulo remissiore, idem mansit Cœli vultus; Juxta Solis ortum, ad instrumenta me contuli, languente jam omne spe observationis habendæ, tuboque 24 pedum in plagam Solis verso, patienter expectavi, an per Nubium aliquem hiatum conspici possit desideratissimus Phœbus: Juxta horam octavam Nubes rarescere ceperunt, ita ut 8 *h*. 36 *m*. Sole clare conspecto, Mercurium nondum intrasse pronunciavi; inde brevibus intervallis sæpius eluxit, ac sequentem habui observationem.

A In

The appendix of *Catalogus Stellarum Australium* in which Edmond Halley described his observations of a transit of Mercury. Collection: Powerhouse Museum, Sydney

Halley then went on to calculate the form of the next expected transit of Venus in 1761 and even to propose suitable places for observation, such as Bencoolen in India and Hudson's Bay in Canada, as these would have a relatively large difference in transit duration. Unfortunately, in Halley's time the orbit, or path, of Venus around the Sun had not yet been fully worked out so his calculations for the 1761 transit were not quite right. Still, the basic method had been established and communicated to interested astronomers.

The 1761 transit of Venus would be nearly half a century after Halley published his paper and his exhortations to observe it were addressed in the manner of a biblical prophet to astronomers of a future generation:

> I recommend it therefore again and again to those curious astronomers who (when I am dead) will have an opportunity of observing these things, that they would remember this my admonition, and diligently apply themselves with all their might in making this observation, and I earnestly wish them all imaginable success: in the first place, that they may not by the unseasonable obscurity of a cloudy sky be deprived of this most desirable sight, and then, that having ascertained with more exactness the magnitudes of the planetary orbits, it may redound to their immortal fame and glory.

FROZEN PLAINS AND TROPICAL SEAS

1761

The 1639 transit had been witnessed by only two observers. The next one, in June 1761, would be a complete contrast. In the years leading up to it, over 100 teams of astronomers in Europe began to make preparations to observe it from a variety of far-flung places. This piece of astronomical history would not be allowed to pass unnoticed.

Cape Town . from Table Bay

It took over a century of efforts by the finest minds before the problem of longitude was solved.

ESTABLISHING YOUR POSITION, whether on sea or land, depends on two coordinates: latitude – the distance north or south of the equator – and longitude – the distance in degrees east or west of any predefined point on the globe, generally taken to be Greenwich Observatory in London. Latitude can be found relatively easily by measuring the angular height of the Sun during the day or the stars at night. Finding longitude is much more difficult.

Longitude can be found by using the 24-hour spin of the Earth. If the time is known at a fixed place (such as Greenwich), navigators can compare it with the local time and establish how many hours the Earth will need to turn before the local time is the same as it was in Greenwich at the time of observation to calculate their longitude. That may sound simple in theory, but in practice it took over a century of efforts by the finest minds before the problem of longitude was solved.

Early navigators had to rely on dead reckoning to estimate how far they had sailed east or west. Dead reckoning involves determining the present location of a ship by taking into account its course and estimated speed from a previous location. During the 17th century, this unreliable method led to a number of ships of the Dutch East India Company accidentally reaching the west coast of Australia. The company had instructed its captains to sail east from the Cape of Good Hope making use of the winds of the 'Roaring Forties' for about 7000 kilometres and then turn north towards Batavia, present-day Jakarta in Indonesia. However, the captains had no way of knowing how far they had travelled and some ships ended up bumping into the land they named New Holland instead.

In 1612, the Italian scientist Galileo observed one of the four moons of Jupiter he had discovered

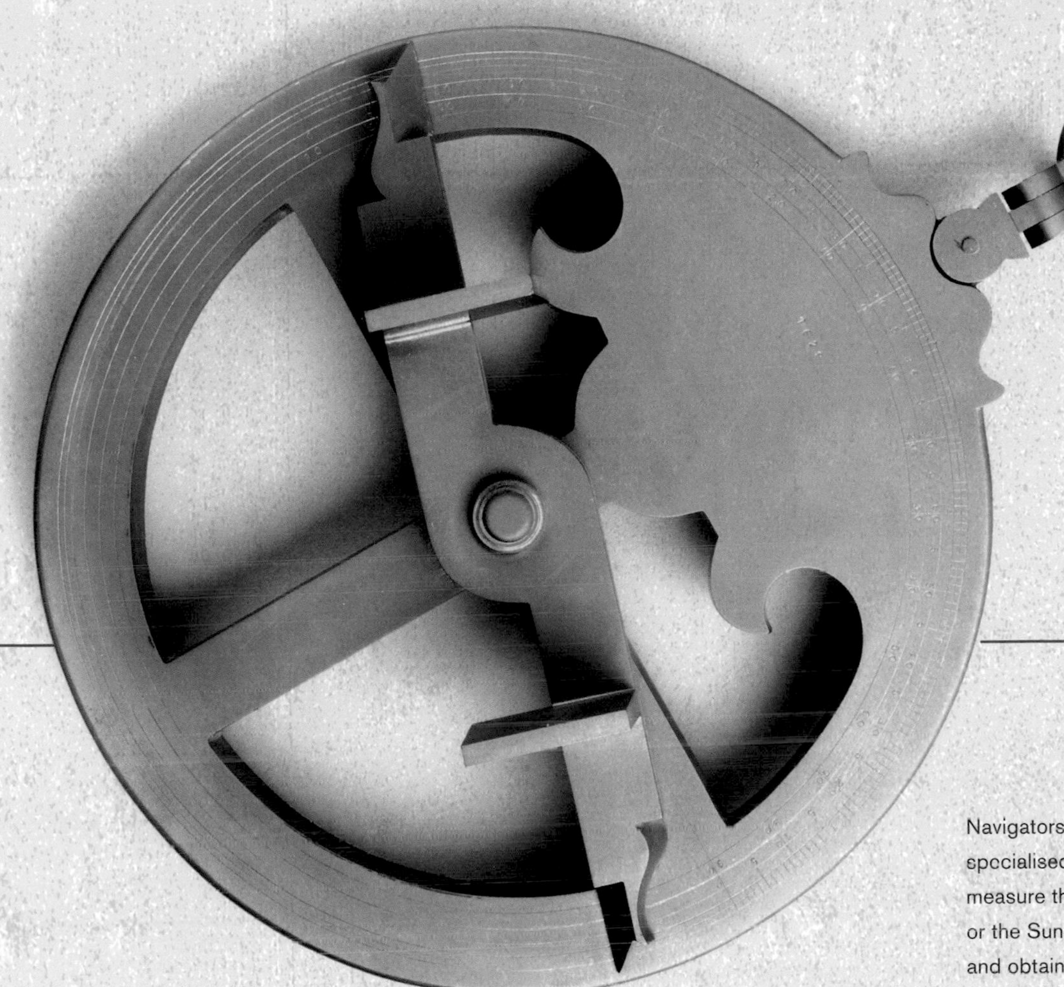

Navigators needed specialised instruments to measure the altitude of stars or the Sun from the horizon and obtain latitude. One of these was the mariner's astrolabe, which was made of heavy brass so that it would hang vertically when suspended on a rope. The holes in the instrument minimised movement by the wind. Collection: Powerhouse Museum, Sydney. Photo: Sotha Bourn

go into eclipse, that is, disappear into the shadow of the planet. Galileo realised that if such events could be accurately predicted they would act as a clock suitable for finding longitude. Over half a century later, the French astronomer Gian Domenico Cassini produced tables of suitable accuracy, but it was found to be too difficult to observe the moons from on board a ship buffeted by waves. For land-based surveyors, though, the method using Jupiter's moons turned out to be extremely useful.

The solution for shipboard navigation did not come until the 1760s when two competing methods became available

Cook used the method of lunar distances and became the first navigator to know his position at all times.

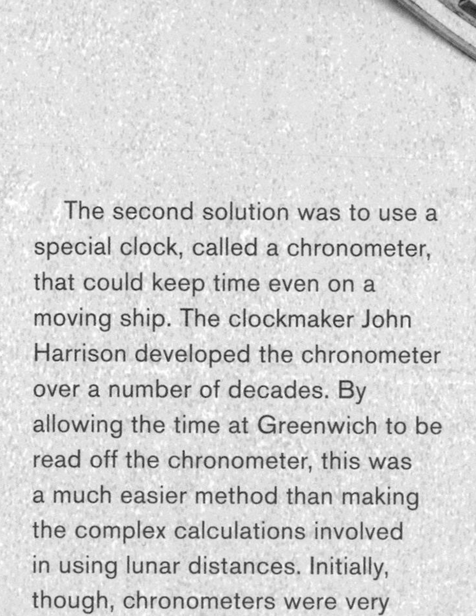

more or less at the same time. The method promoted by England's Astronomer Royal, Nevil Maskelyne, was to find the time at Greenwich by using the motion of the Moon as a clock. To simplify such calculations, from 1767 onwards Maskelyne published the annual *Nautical Almanac and Astronomical Ephemeris* with tables listing the angular distances of bright stars from the Moon at different times. Equipped with *Nautical Almanacs*, Captain James Cook on his first and most famous voyage (1768–71) used the method of lunar distances (lunars) to determine longitude and became the first navigator to know his position at all times.

The second solution was to use a special clock, called a chronometer, that could keep time even on a moving ship. The clockmaker John Harrison developed the chronometer over a number of decades. By allowing the time at Greenwich to be read off the chronometer, this was a much easier method than making the complex calculations involved in using lunar distances. Initially, though, chronometers were very expensive and could only be sent on a few ships. On his second voyage (1772–75) Captain Cook carried a replica of Harrison's successful chronometer built by the clockmaker Larcum Kendall for the then large sum of £400. By the end of the 1700s, chronometers were being produced on a large scale at much lower prices.

< A special angle-measuring device called a sextant was developed to allow ships' navigators to measure the angle between stars and the Moon for the method of lunar distances. This sextant was no. 1787 among those built by the English instrument maker Matthew Berge. Collection: Powerhouse Museum, Sydney. Photo: Sotha Bourn

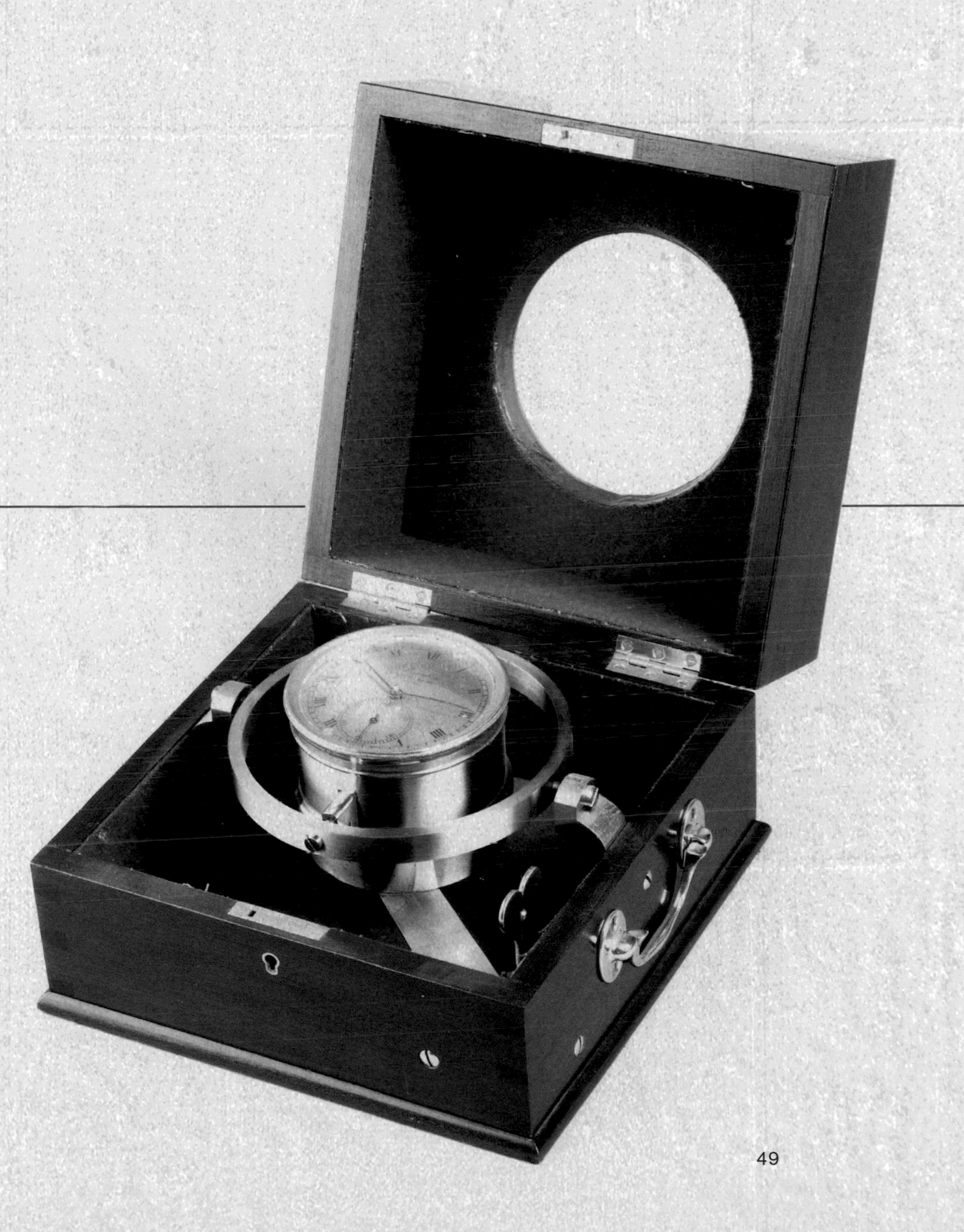

> In the late 1700s, John Arnold and his rival Thomas Earnshaw developed chronometer designs that were easier and cheaper to make than the designs of John Harrison. This chronometer, built by Thomas Earnshaw in 1801, was used by the famous navigator Matthew Flinders on board HMS *Investigator*. Collection: Powerhouse Museum, Sydney. Photo: Sotha Bourn

In practice, accurately timing the transit of Venus was not as easy as Edmond Halley had predicted it would be 50 years earlier. One obvious problem was that Halley's method required both the beginning and end of the transit to be visible from each observing site. This limited the possible sites and, of course, there was also a risk that the sky might not be clear at both the beginning and end of the transit.

To address these problems, the French astronomer Joseph-Nicolas Delisle modified Halley's method so that it was only necessary to observe Venus either entering or leaving the disc of the Sun. In Delisle's method, it was the actual times at the beginning or the end of the transit that had to be compared. Today, with GPS receivers and accurate clocks that would be ridiculously easy, but it was not so in the 18th century when techniques for finding position on the Earth's surface were still being developed. Accurate comparison of the times at different places, especially in places far from Europe, were difficult and sometimes impossible. As a result, astronomers employed a mixture of both Halley's and Delisle's methods when observing the transits of the 18th and 19th centuries.

In the background, as at the time of the previous transit, war raged. What became known as the Seven Years' War broke out in 1756 and lasted until 1763. Like the Thirty Years' War, its causes were complex, but this time they were mainly about trade and colonial expansion by the European powers. Hence the bitter conflict was not only fought on European soil, but elsewhere on the globe, including India and America.

By the time of the war, Britain had established a number of colonies along the Atlantic coast of North America with a population of around a million and a half. The French also had territories there and Britain took the opportunity provided by the war of attacking them. Despite many of the Native American tribes supporting the French, the British managed to generally prevail. It was, however, the elimination of the French threat that allowed some of the colonies a decade or so after the end of the conflict to try to break away from British control in the American War of Independence.

Culture was changing at this time too. The five-year-old child prodigy Wolfgang Amadeus Mozart and his equally talented sister made their first performances before royalty in Munich, Vienna and Prague in 1762. In England a few years earlier, the author Samuel Johnson published his *Dictionary of the English Language* to much acclaim, while the portraits of the painters Sir Joshua Reynolds and

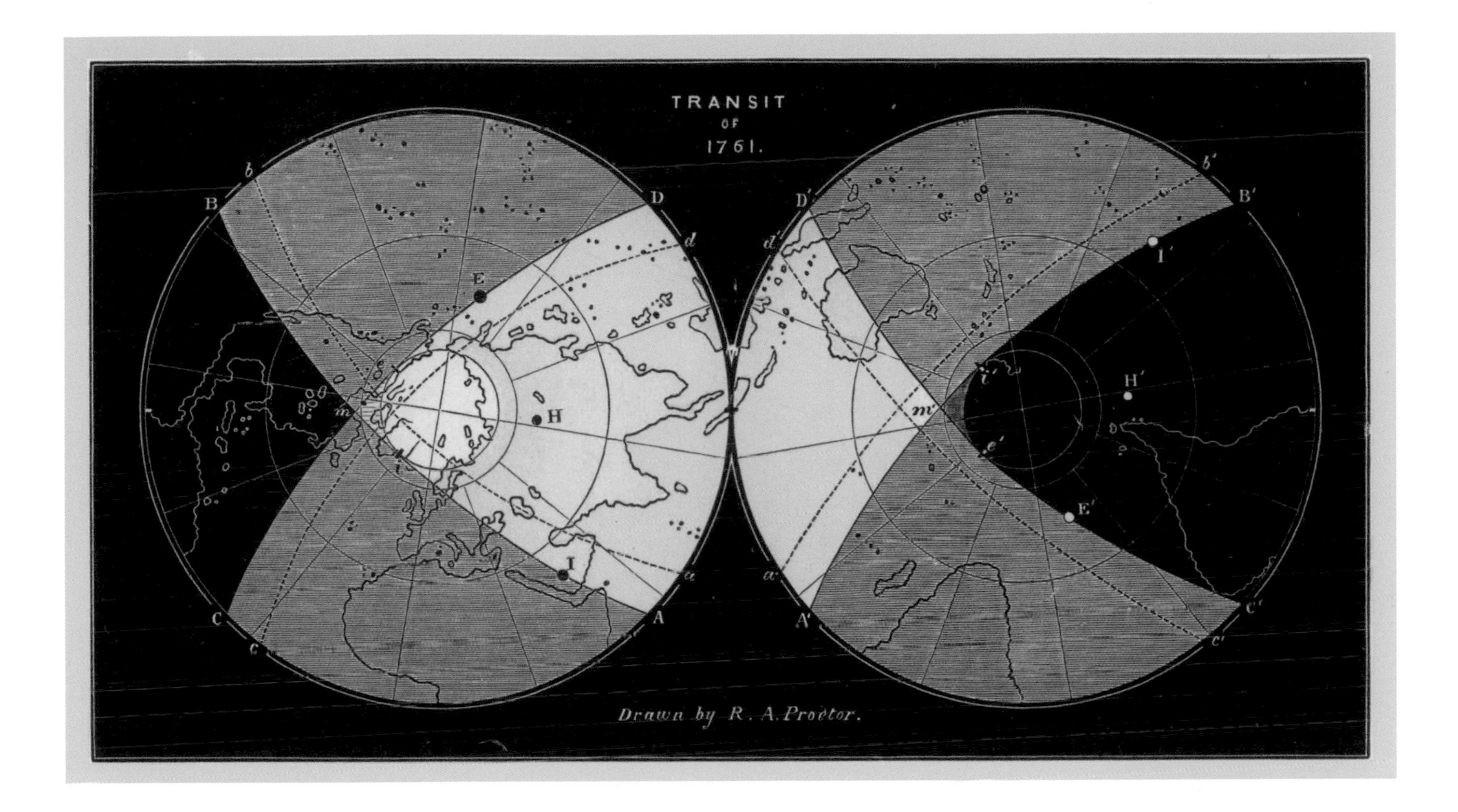

The visibility of the 1761 transit of Venus, with the northern hemisphere on the left and the southern hemisphere on the right. The brightest area denotes the regions from which the transit was fully visible, the shaded areas indicate regions from which the beginning or end of the transit was visible, and the darkest areas are the regions from which the transit could not be seen. Map from *Transits of Venus* by RA Proctor, 1874. Brian Greig collection.

Thomas Gainsborough were becoming better known.

Back in 1686 and 1687, Sir Isaac Newton had put astronomy on a more scientific basis with the publication of the *Principia*. In this work he develops his laws of motion for moving objects and the universal law of gravity and from them explains the motions of the planets around the Sun. A more practical development in astronomy took place closer to the time of the 18th-century transits: a major improvement in lens telescopes. From about 1758 the optician John Dollond began selling telescopes with achromatic lenses that brought different colours to the one focus. Equipped with

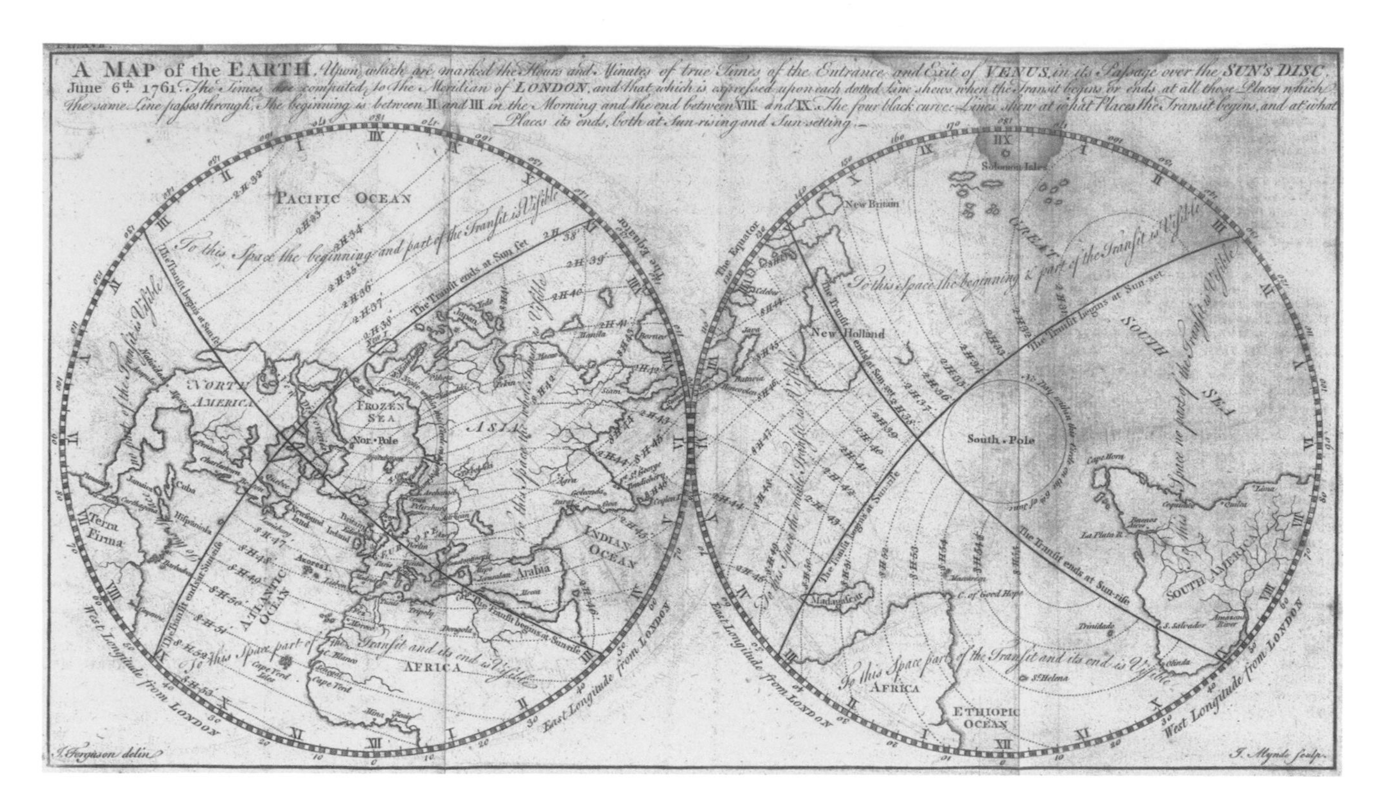

such lenses, smaller and higher quality telescopes providing sharp images became available, at least to those who could afford them.

Another, even more important development that exactly coincided with the two 18th-century transits was the solution of the problem of longitude (see box, pages 46–49). To determine position on land or at sea involves finding two coordinates, latitude and longitude. The first of these is relatively easy to find, but for shipboard navigators finding the latter coordinate was extremely difficult. In the 1760s two competing methods of finding longitude became available almost simultaneously. One, based on astronomical observations of the Moon, was the method used by James Cook on his voyage to Tahiti to observe the transit of 1769 (see the next chapter), while the other method was based on an accurate clock – a chronometer – that would work at sea.

∧ The Scottish astronomer James Ferguson redrew Delisle's map of the world with a polar view, showing the northern and southern hemispheres. Brian Greig collection

> The *mappe monde*, or complete world map, made by Joseph-Nicolas Delisle in 1760 in preparation for the 1761 transit of Venus. British Library, King George III Topographical Collection, Maps K.Top.1.77

THE FIRST GLOBAL SCIENTIFIC EVENT

Halley's exhortation of 1716 to observe the transits of Venus was enthusiastically taken up for the transit of 1761. According to the 19th-century astronomy writer Richard Anthony Proctor, 176 astronomers took part in observing the transit from 117 locations. With astronomers from many countries taking part in the observing campaign, this was the first global scientific event.

Most of the participants made their observations from their home observatories, but a few undertook extensive travels to the far corners of the globe to be in suitably advantageous locations on the day of the transit. Travel in the middle of the 18th century was unimaginably more difficult than it is today; it involved many months of hardship travelling by ship or on land. To complicate matters even further, between 1756 and 1763, a period that encompassed the transit and the advance preparation that was

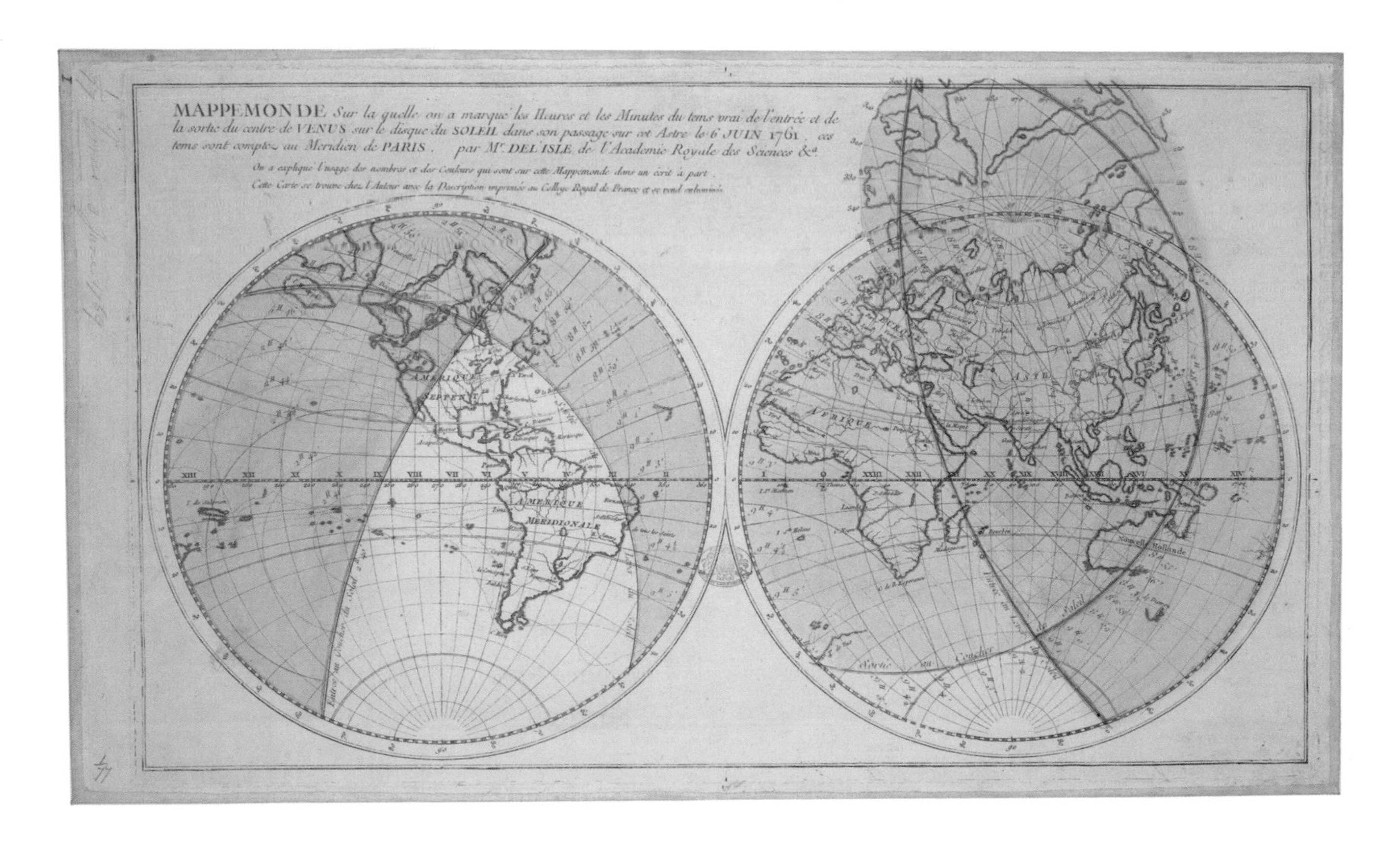

required, Britain and France, together with their respective allies, were engaged in the bitter struggles of the Seven Years' War.

In this chapter we will relate the adventures of two Englishmen, Charles Mason and Jeremiah Dixon, who subsequently became famous in the United States in connection with the Mason–Dixon Line. As well, we will look at the separate adventures of two Frenchmen, Guillaume Le Gentil and Abbé Chappe d'Auteroche.

Some of the distant observing locations were chosen because they had been mentioned by Halley back in 1716. However, in Halley's time the path of Venus around the Sun was not yet known to sufficient accuracy, which meant that some of the observing stations he had suggested were unsuitable. The key task was to find locations from which the transit was visible from beginning to end so that the duration of the transit could be timed. Then the next task was to select from these possible sites places where the difference in durations was as large as possible. It turned out that in 1761 there was little difference in the expected durations between different sites, however far they were apart geographically.

In the French astronomer Joseph-Nicolas Delisle's alternative method, just one observation of the transit, either of Venus entering on to the disc of the Sun, called ingress, or of Venus leaving the disc of the Sun, called egress, was sufficient to yield results. Thus a site could be useful even if only the beginning of the transit was visible before sunset or the end of the transit was visible after sunrise. What was needed was to make observations from a place where, say, ingress was in advance of other places and from a place where ingress was delayed with respect to other places. To simplify site selection Delisle published a *mappe monde*, a map of the globe divided into two hemispheres, with lines indicating where ingress and egress were to be visible.

With Delisle's method the times of ingress or egress were to be compared between sites. This meant that the observers had not only to time the instant of ingress or egress according to local time, but they also had to determine their longitude, how far east or west of Greenwich or Paris they were, to a good degree of accuracy. With those two determinations the times of ingress or egress could be reduced to times at a standard location such as Greenwich so that times from different locations could be compared. This requirement was to be the major handicap of Delisle's method, since in the 18th century the accurate determination of longitude was a challenging task for astronomers, even on land.

Map of the island of St Helena prepared by Jacques Nicolas Bellin and published in Paris in 1764. National Library of Australia, nla.map-rm324-e

A view of Jamestown on St Helena, engraved from a drawing by ES Blake in 1830. St Helena Virtual Library and Archive, <www.bweaver.nom.sh>

TRIALS AND TRIBULATIONS: THE BRITISH EXPEDITIONS

Encouraged by the Royal Society – a distinguished assembly of those interested in all aspects of science – and with initial funding of £1600 provided by King George II, Britain decided to send out two expeditions to observe the transit from remote locations. One of these was led by the Reverend Nevil Maskelyne, who was depicted, somewhat unfairly, as the villain in the story of John Harrison and the development of the chronometer in Dava Sobel's 1995 book *Longitude*. Maskelyne, accompanied by Robert Waddington as his assistant, was to go to the island of St Helena in the South Atlantic Ocean, the same island from which Edmond Halley a century earlier had observed the transit of Mercury. (Just over 50 years later, it became the final place of exile for the deposed French emperor Napoleon Bonaparte.)

In 1836 the naturalist Charles Darwin visited the island near the end of the second voyage of the *Beagle*. Darwin gave the following dramatic description of St Helena:

> This island, the forbidding aspect of which has been so often described, rises abruptly like a huge black castle from the ocean. Near the town, as if to complete nature's defence, small forts and guns fill up every gap in the rugged rocks. The town runs up

Fort Munden on St Helena painted by Lieutenant William Innes Pocock in the early 1800s. © National Maritime Museum, Greenwich, London

a flat and narrow valley; the houses look respectable, and are interspersed with a very few green trees. When approaching the anchorage there was one striking view: an irregular castle perched on the summit of a lofty hill, and surrounded by a few scattered fir-trees, boldly projected against the sky.

On 6 April 1761 Maskelyne and Waddington arrived at James Fort (now Jamestown) on the British East

India Company ship *Prince Henry*. By 26 May the administration of the island was writing back to London that, 'We have already erected an observatory for them in the country and shall furnish whatever else the service may require'. Sadly, when the transit took place 11 days later on 6 June the weather was cloudy and Maskelyne was unable to make observations in spite of careful preparation.

The other British expedition was more successful and involved much more adventure than Maskelyne's. Its two members were Charles Mason, who had been an assistant observer at Greenwich Observatory, and the surveyor Jeremiah Dixon. Their work subsequent to the transit in surveying the border between the colonies of Maryland and Pennsylvania in North America has made them household names, especially in the United States.

Charles Mason was born in April 1728 in Gloucestershire, England, and was the son of a baker and miller. He attended the Tetbury Grammar School, but also received additional schooling in mathematics. The fortunate circumstance of living near the Astronomer Royal, James Bradley, secured Mason a position as assistant at Greenwich at the age of 28. There he was involved in compiling tables of lunar distances, the angular distances of bright stars from the Moon. These were to be used by navigators on board ships at sea to determine longitude.

Jeremiah Dixon, born on 27 July 1733 in County Durham, England, came from a reasonably wealthy family (his father owned a coalmine). After attending a local school, he became a surveyor, though nothing is known about his training and he may have been self-taught. When asked where he had learnt his astronomy, he made the cryptic remark that it was 'In a pit cabin on Cockfield Fell', a local hillside. Like Mason he made a fortuitous local connection; in his case it was with the instrument maker John Bird. It is believed to be Bird who recommended Dixon as Mason's assistant for the transit of Venus observations.

Mason and Dixon were to observe the transit from the British settlement of Bencoolen in Sumatra (now the Indonesian city of Bengkulu). On 8 December 1760 Charles Mason reported to the Royal Society that they were on board HMS *Seahorse*, a Royal Navy frigate under the command of Captain James Smith, and ready to sail. However, three days after leaving Portsmouth, on 8 January 1760, the *Seahorse* was involved in a serious skirmish lasting over an hour with a French warship, *Le Grand*. When another Royal Navy ship appeared, the French vessel escaped and the *Seahorse* limped back to port with 11 dead and 38 wounded among the crew. Not surprisingly, after this incident the two observers became reluctant to embark

A plan of the border between Pennsylvania and Maryland, the Mason–Dixon line. Only a narrow strip on either side of the border was surveyed and is indicated on the plan. The surveyors, Charles Mason and Jeremiah Dixon, observed the 1761 transit from Cape Town, South Africa. Published by Robert Kennedy, Philadelphia, 1768. Library of Congress Geography and Map Division, Washington, DC

again on such a dangerous voyage and Mason wrote to the Royal Society that 'We will not proceed thither, let the Consequence be what it will'.

These comments did not please the Royal Society. It insisted that the two men depart, and told them in the strongest possible terms that their refusal, 'cannot fail to bring an indelible Scandal upon their Character, and probably end in their utter Ruin'. In case those comments were not severe enough, the letter went on to threaten to take the two observers to court. Mason and Dixon salvaged whatever dignity was left to them and sent a brief reply that they would sail that evening.

The *Seahorse*, commanded this time by a different captain, sailed again on 4 February. By now it was too late to reach Bencoolen in time, so instead the ship headed for the Dutch colony of Cape Town on the southern tip of Africa. On arrival at Table Bay on 27 April, Mason and Dixon asked the local authorities for permission to stay and for help in setting up an observatory. They were given excellent assistance and provided with a place for their observatory near what is now the centre of Cape Town, a spot between St Johns Road and Hope Street and behind the later St Mary's Cathedral.

Their observatory was circular with a width of 4 metres and a height of 1.7 metres. Of course, they could not put a hemispherical dome on top, as in a modern observatory, but instead had a conical roof that could be turned by hand so that its opening could face towards any part of the sky. To furnish the observatory Mason and Dixon had brought a number of instruments with them from London. There was an astronomical quadrant made by the famous London instrument maker John Bird and lent to them by the president of the Royal Society, the Earl of Macclesfield. They also had two reflecting telescopes of 2-feet (60-centimetre) length by the eminent Scottish-born telescope maker James Short, and a pendulum clock, which they secured to two solid timber posts sunk about a metre into the ground.

Their first task was to check out the rate of the clock, that is, how much time it was losing or gaining each day. Observations of the bright star Procyon with the quadrant allowed them to establish that the clock was losing about 2 minutes 17 seconds a day. Note that the error of a clock is of no concern to astronomers or navigators as long as the error is known and consistent.

On the day of the transit, Mason and Dixon met the astronomer's great and everlasting enemy, clouds. Fortunately, there were clear periods during which they could observe the Sun and Venus. They made a series of measurements of the position of the planet on the Sun's disc and the relative apparent sizes of the two bodies. Most importantly, they could time the moment Venus was about to move off the Sun, an instant known as the internal contact at egress, or third contact. In his report Mason stated, 'When I saw the planet first its periphery and that of the Sun's were in a great tremor, but this vanished as the Sun rose and became well defined. Four minutes before the internal contact the Sun's disc was entirely hidden by cloud for about one minute'.

To make their measurements useful, especially the timing of the egress, Mason and Dixon had to carefully establish the geographical coordinates of their observing site. To do this they stayed on at Cape Town for nearly four months after the transit making observations of stars with the Bird quadrant. They augmented these observations by observing Jupiter's satellites with the telescopes as events associated with those gave another measure of time at Greenwich and hence longitude.

Mason and Dixon did not leave directly for England, but sailed on 3 October 1761 aboard the astronomically named ship *Mercury* for St Helena. There they assisted Maskelyne with his observations. In particular they compared the daily rate of the clock they had used at Cape Town with its rate on the island. Given a pendulum of constant length, the rate of the clock is affected by the local strength of gravity. The Earth is not a perfect sphere, so gravity varies slightly from place to place and these timing measurements helped Maskelyne in his project to define the shape of the Earth.

Around the end of 1761 the two astronomers who had been at Cape Town, together with Maskelyne and his assistant, sailed for home and reached it safely. Mason and Dixon returned with the knowledge that they had carried out fine and professional work on the transit of Venus. Recognition of their expert partnership led to their selection to measure the disputed boundary between the American colonies of Pennsylvania and Maryland, a task that took them almost four years, from 1763 to 1767.

In the United States, the Mason–Dixon Line – as the adjusted boundary became known – later came to stand for the division between the 'free' northern states and the slave-owning southern states, a division that would lead to civil war in the 1860s. And 'Dixie' became a slang term for the southern states, giving rise to the name Dixieland for the type of early jazz music that originated in New Orleans.

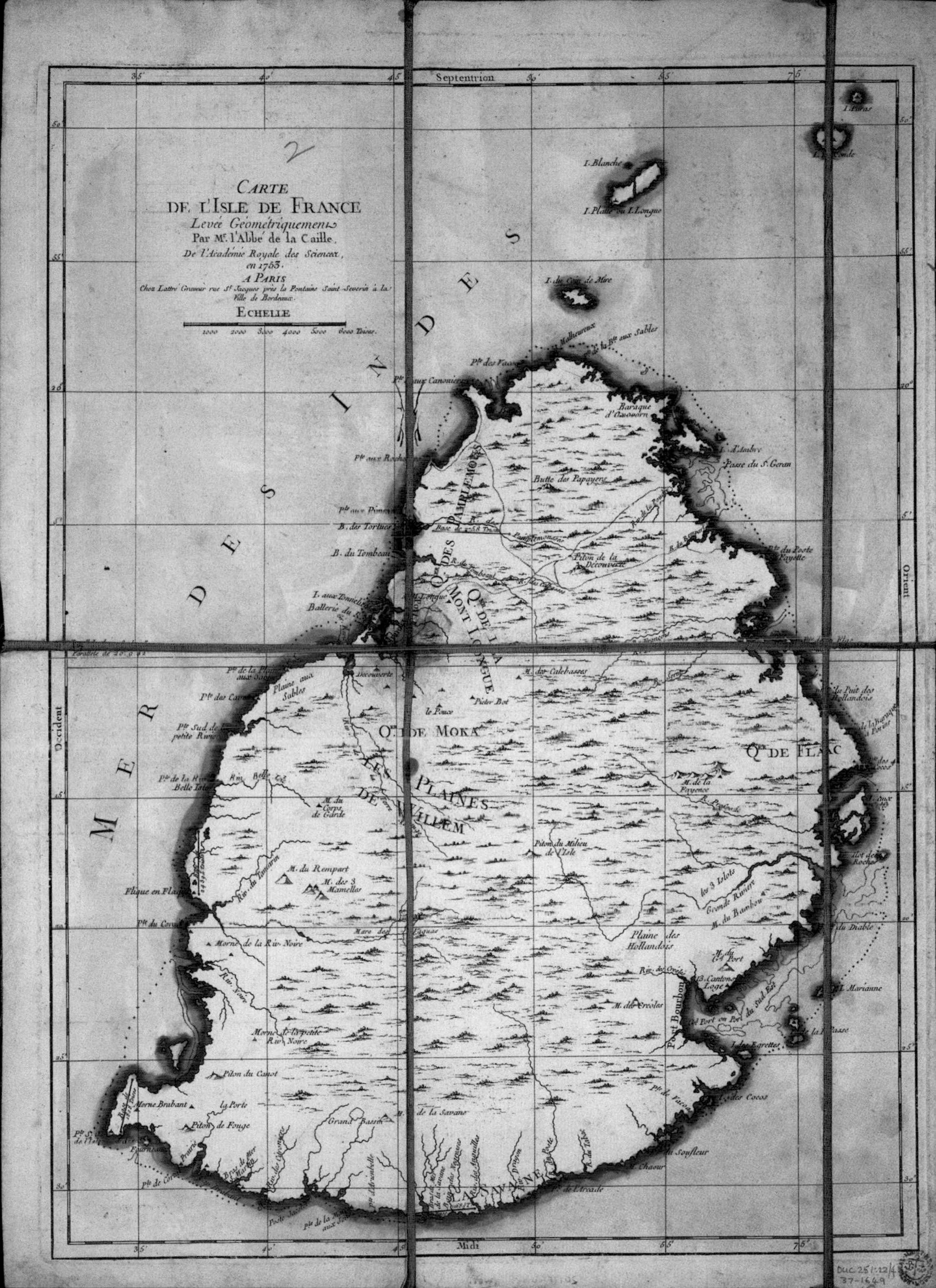

A map of the Isle de France (Mauritius) prepared by Abbé Nicolas Louis de Lacaille in 1753. Lacaille is important in the history of southern hemisphere astronomy, for he prepared the first catalogue of southern stars after that of Halley in the previous century and delineated and named 14 constellations that we still use today.

CARTE
DE L'ISLE DE FRANCE
Levée Géometriquement
Par Mr. l'Abbé de la Caille,
De l'Académie Royale des Sciences,
en 1753.
A PARIS
Chez Lattré Graveur rue St. Jacques près la Fontaine Saint-Severin à la Ville de Bordeaux.
ECHELLE
1000 2000 3000 4000 5000 6000 Toises.

Mason and Dixon returned to England from America in time to observe the 1769 transit of Venus. This time Mason observed from Ireland while Dixon travelled to Norway. The latter died in 1779. Mason continued to work on tables of the Moon and the *Nautical Almanac*. Near the end of his life Charles Mason returned with his family to Philadelphia, where he died just a few months after arriving, on 25 October 1786.

IN THE WILDS: THE FRENCH EXPEDITIONS

Let us now turn to the tale of the unlucky French scientist with the glorious name of Guillaume Joseph Hyacinthe Jean Baptiste Le Gentil de la Galaisière, whose transit expedition was the longest in history. Le Gentil was born on 12 September 1725 in the town of Coutances in the western part of Normandy, 65 kilometres to the south of Cherbourg. While studying theology in Paris he came under the influence of Joseph-Nicolas Delisle, a professor at the Collège Royal, later to become the Collège de France. Though Le Gentil became an abbé, he was more interested in astronomy than religion and took to visiting Paris Observatory. He willingly accepted an offer to work there and was so successful that by the age of 28 he was a member of the French Academy of Sciences.

When the Academy began planning expeditions for the 1761 transit of Venus, Le Gentil volunteered to take part. He was selected to make the observations from the French colony of Pondicherry on the southeast coast of India.

The first stage in Le Gentil's journey to India was to sail around the Cape of Good Hope to Mauritius, then known as Îsle de France (see p. 63). He left France aboard the French East India company ship the *Berryer* on 26 March 1760 and reached Îsle de France on 10 July. From there he planned to take passage on a ship bound for India, but no ship was available, and in any case war had broken out between France and Britain making such voyages difficult and dangerous. To compound his troubles, while waiting to travel, Le Gentil became ill with dysentery.

Le Gentil's circumstances improved with the decision by the island authorities to send a naval frigate, the *Sylphide*, to India. The ship sailed on 11 March 1761 and though the date of the transit in June was getting perilously close, Le Gentil was assured that the *Sylphide* was a fast vessel that would reach its destination in two months. It was not to be. When the ship reached India after a slow voyage during which contrary winds blew the ship the wrong way, he and the ship's company were informed that Pondicherry had fallen to the British.

Sleds used for travelling in Russia, plate II from Chappe d'Auteroche's book, *A Journey into Siberia, Made by Order of the King of France.* National Library of Australia

The return voyage after departing on 30 May was much faster than the outward one. Yet it was not fast enough for Le Gentil as he found himself at sea on the day of the transit on 6 June. Though he had an excellent view of both the beginning and end of the event, scientifically his observations were of no use because he could neither do exact timing of his observations nor determine his longitude and hence the Paris time to sufficient accuracy.

Frustrated and disappointed by his lack of success in making meaningful observations of the transit, Le Gentil decided to stay for a while at Mauritius and explore and map the surrounding area. He visited the much larger island of Madagascar, starting with the French settlement of Fort Dauphin. There he contracted an illness that he says was like a 'violent stroke', which was treated with the then usual medical method of blood-letting. He was left for a while with double vision, though we cannot tell whether that was from the disease or the cure.

After four years of map-making and studying the area, Le Gentil began thinking of not returning to France, but instead of waiting for the 1769 transit. We will continue with his adventures in the next chapter.

Another brave French astronomer who ventured to distant lands to make his observations was Abbé Jean-Baptiste Chappe d'Auteroche, who set out to observe from Tobolsk in Siberia. Tobolsk was selected as an observing site because from there the transit could be seen from beginning to end, and the duration would be shorter than anywhere else in the world. This was important since better results for the distance could be obtained if there were large differences in duration between observing sites. Chappe published a book describing his travels and his careful observations of people and customs along the way, so we know a good deal about his trip. Born in 1728, Chappe seems to have come from a fairly important family as one of his older relatives, Pierre Chappe, was an adviser to the French King, while his brother Ignace-Urbain Chappe became a lawyer in Parliament.

As to be expected, without the modern roads, cars, railways and other conveniences we now take for granted, he had a difficult journey, with many mishaps along the way. Contrary to what we may expect, though, as we will see, travel in Siberia at that time was easier and faster in the depth of winter than at other times. Chappe set out at the end of November 1760 and soon encountered his first problem. On the road from Paris to Strasburg all his barometers and thermometers broke when one of the carriages fell into a ditch.

To try to avoid further breakages, he took a boat along the Danube to Vienna, disregarding a warning that at that time of the year fogs increased the hazards of the journey. On this boat trip he helped to save a young man who was about to kill himself after a quarrel with his girlfriend. At another time someone stole one of Chappe's bags containing most of his clothes while he was ashore at a village Christmas Eve church service.

On 31 December 1760 Chappe reached Vienna, where he was well received by the Emperor and Empress. From there he continued by road to Warsaw and Poland. He makes the interesting observation that people travelling in that country to stay with friends need to take with them all common necessities, such as tables, chairs and beds, since each family only had enough for themselves.

By the time Chappe reached Riga, then in the Russian Empire but now capital of independent Latvia, the ground was covered in snow and he had sledges placed under the carriage wheels. Unfortunately, on leaving the town they soon reached a plain without snow, making the sledges useless. By this time, it was night. At first, the coachmen refused to seek help to remove the sledges, but they eventually did so, on payment of a rouble. For two roubles, a sum that Chappe felt was too high, some local people removed the sledges from under the wheels of their overloaded coaches so that Chappe and his party could continue. But the snow soon returned, the carriages had to be abandoned and the journey to St Petersburg continued on sledges.

On 13 February 1761 Chappe reached St Petersburg, at that time the capital of the Russian Empire. There he found that the Russian astronomers planning to observe the transit from Siberia had left a month ago. With time growing short, the local scientists suggested that he go to a nearer and more accessible place elsewhere in Russia. Chappe, however, was adamant that Tobolsk was the most advantageous location and that that was where he was going.

For the journey to Siberia he had to take with him all the necessities, from bread to beds, because there was no expectation of obtaining them on the way. He hired an interpreter, as well as a watchmaker, who could repair the clocks in the likely event that they were damaged on the journey. In those days, long before the construction of the Trans-Siberian Railway, travel was actually quicker in winter because sledges could easily glide on snow. Chappe's worry was that it was already late in the winter season and that the snow would thaw.

He set out from St Petersburg on the evening of 10 March 1761 with four sledges: a covered one drawn by five horses for Chappe himself; the watchmaker and the interpreter in a half-covered

one; one for a 'sergeant' who was sent along as a guide with the provisions; and one for Chappe's instruments. The first stop was Moscow. On the way there, Chappe faced difficulties with his companions. The watchmaker wanted his own sledge, which was refused as an unnecessary expense at a time when he and the others were quickly using up the provisions, especially the liquid ones.

At Moscow, Chappe had to replace the sledges, which were already damaged beyond easy repair by the shocks of the four-day journey between Moscow and St Petersburg. He also replenished his supplies and after departure informed the watchmaker and the interpreter that they would be left along the road if they wasted time on the way.

The sledges went quickly on the snow and even more quickly over frozen rivers. Speed led to a near disaster when one of the horses fell into an unfrozen opening in the thick ice covering a river. Chappe and his companions managed to prevent the unfortunate animal from dragging the other horses and a sledge into the hole as well by the expedient of cutting the stays that tied the horses to the sledge. In spite of this mishap, where possible they travelled on rivers, including a long stretch on the Volga. To enjoy the speed of travel Chappe sometimes left the covered part of his sledge and positioned himself behind it. He could not last long outside, however, as the thermometer was down to 22°C below zero.

As Chappe was approaching Tobolsk, the thaw was starting to set in. At a place 100 kilometres or so from his destination he needed fresh horses to cross the Tobol River, but the locals refused on the grounds that it was too dangerous to cross. Chappe succeeded in overcoming these objections with some fast thinking and the use of his thermometer, a device that was as yet unknown there. He had taken the thermometer inside the over-heated house where the negotiations were taking place. Noticing the locals' interest in the instrument he explained that when taken outside the mercury would drop to a much lower value. Pointing to a value just below zero degrees, he told them that if it falls to that point or lower it would be safe to cross the river. The locals took the instrument outside and soon 'the most stubborn fellow' among them reported that 'the animal had got below the mark'. After that Chappe's only problem was to prevent the interpreter explaining that mercury was not an animal.

Chappe finally reached Tobolsk on 10 April 1761 after travelling for more than 2000 kilometres by sledge from St Petersburg. There, with the support of the governor of the city, he found a spot for his observatory a kilometre or so outside the town and

TRANSIT OF VENUS

The Irtysh River, Tobolsk's old city and Kremlin as it looks today. Photo: Thomas Claveirole

on a hilltop to provide a good view to the horizon. Once the observatory was built, probably with prefabricated components brought from Paris, Chappe installed his quadrant, clocks, a telescope and other instruments. Telescopes at that time generally had a long focal length to reduce the problem that simple lenses could not provide sharp focus for different colours; hence Chappe's telescope was about 6 metres long.

The local people, with the exception of the governor and a few other leaders, could not make any sense of Chappe's mysterious observatory, strange instruments and nocturnal activities and decided that he was a magician. By itself that belief would not have been of concern, but that year as the snow melted, the local river, the Irtysh, overflowed its banks to a greater extent than in previous years and flooded part of the town, drowning some people. The inhabitants, of course, blamed the 'magician' for this misfortune. To secure his own safety and that of the observatory, Chappe spent his nights there accompanied by a guard of three soldiers and their sergeant.

Prior to the transit, Chappe busied himself with observations to establish his longitude and to test the equipment. In his account of the expedition, he explains that he could not establish longitude by the usual astronomical technique of observing the moons

Tobolsk and the River Irtysh in 1912. Library of Congress, Prints & Photographs Division, Prokudin-Gorskii Collection, LC-DIG-prokc-20755

of Jupiter because the sky never became dark enough in the Siberian summer. Fortunately, there was a partial eclipse of the Sun on 3 June and subsequent comparison of his observations with those of a colleague in Sweden allowed an accurate longitude to be established.

On the night before the transit Chappe was greatly distressed by the appearance of fog and clouds. By morning, though, these began to disperse and all seemed well. For the much-awaited event Chappe was joined by local dignitaries: the governor, an aristocrat and the archbishop with some of his monks. To avoid being disturbed in his tasks Chappe provided these people with their own tent and telescope. He was also worried about being inundated by the local inhabitants, but discovered that they were all sheltering in fear in their homes or in churches.

There were still clouds hovering about the Sun as Venus first touched the disc, but these disappeared as Chappe prepared to time the all-important second contact or ingress. With the help of the watchmaker, who took notes and watched the clock, and the interpreter, who counted the time aloud, Chappe successfully timed the event. He says, 'Pleasures of the like nature may sometimes be experienced; but at this instant, I truly enjoyed that of my observation …'.

With the sky remaining clear Chappe could make observations for the whole duration of the transit, including timing when Venus was about to leave the disc (third contact) and when the planet was completely off the disc. The governor sent these highly precious observations, for which Chappe had exerted so much effort, to the Russian Court by courier. Chappe included two sets of observations: one to the academy at St Petersburg and one to the French Academy in Paris.

His main mission successfully accomplished, Chappe made a slow return through Siberia, making careful observations as he went. Back in Paris he completed his book, *A Journey into Siberia*, published in 1768. This book was well received in France, but in Russia some of Chappe's comments were seen as unfavourable. This led to a chapter-by-chapter refutation in a book titled, *The antidote; or an enquiry into the merits of a book, entitled A journey into Siberia, made in MDCCLXI ... and published by the Abbé Chappe d'Auteroche, ... By a lover of truth*. That lover of truth is believed to have been no less a personage than the Empress of Russia, Catherine II.

It seems that one adventure to view a transit of Venus was not enough for Abbé Chappe d'Auteroche and he travelled to Mexico for the 1769 transit. We will take up that story in the next chapter.

ANALYSING THE RESULTS: 1761

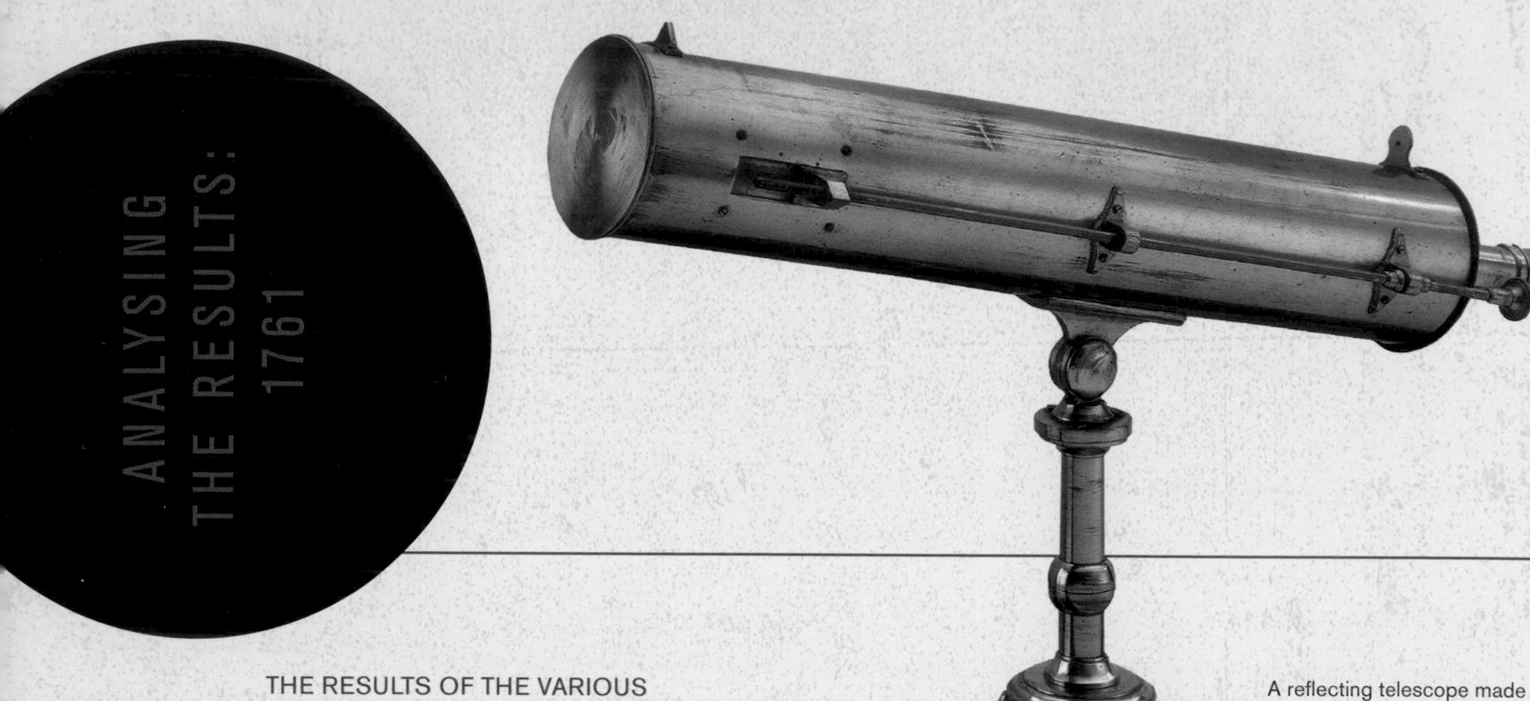

A reflecting telescope made by James Short. © National Maritime Museum, Greenwich, London

THE RESULTS OF THE VARIOUS expeditions and observations of the 1761 transit were analysed by James Short, a Fellow of the Royal Society. Born in Edinburgh, Scotland, in 1710, Short was orphaned at age 10. Encouraged by a professor of mathematics after graduating from Edinburgh University, Short began making reflecting telescopes. He became so successful with these that his fame spread quickly and by 1736 he was summoned to London to teach mathematics to William, Duke of Cumberland, who was the younger son of the King and was later to be known as 'Butcher Cumberland' after the Battle of Culloden in Scotland. Within two years Short had moved permanently to London, where his telescopes commanded twice the price of those of his competitors.

Short observed the 1761 transit from Savile House in Leicester Square, London, which was a royal residence. He was there at the invitation of the Duke of York, who was present for the transit together with a number of assorted royals who were pleased 'to do us the honour of looking at this uncommon appearance'. Short claimed that his instruments were similar to, but slightly better than those at Greenwich and taken by Maskelyne to St Helena and by Mason and Dixon to Cape Town. They included

James Short worked out an average Sun–Earth distance of 152.1 million kilometres, which is not too far from the present-day value.

---

a Shelton astronomical clock and reflecting telescopes with 45- and 60 centimetre focal lengths, probably of his own manufacture. With these he made a number of measurements of the position of Venus during the transit and the time of the internal contact before the planet moved off the Sun's disc.

Short began his analysis with the timings of Venus leaving the disc of the Sun or internal contact at egress from the different locations. After making the best possible estimate of the longitude of each location, he compared Mason and Dixon's timing from the Cape with that of 15 timings from Europe and calculated the Sun's distance or parallax on the day from each. His final value was based on the mean of these, leaving out a few discordant values. From this he worked out an average Sun–Earth distance of 152.1 million kilometres, which is not too far from the present-day value of 149.6 million kilometres.

In addition, Short looked at the observations from places such as Tobolsk, Siberia, where the total duration of the transit was able to be observed. These observations gave a less reliable value because there was less than three minutes' variation in duration between them, with Tobolsk having the shortest duration. Nevertheless, the value calculated from Halley's method of durations was close to the value from the egress timings.

Although Short's value for the distance of the Sun seems a good one, other people also tried to calculate the distance using similar data and obtained a variety of results. Their values ranged from 125.3 to 154.8 million kilometres. This large range was obviously unsatisfactory and astronomers did not regard the observations of the 1761 transit a success.

VENUS OF THE SOUTH SEAS

1769

Sometimes scientific expeditions have unintended consequences. The desirability of observing the 1769 transit from the South Seas began a chain of events that would lead to the founding of the colony of New South Wales by the British in January 1788. In effect, modern Australia owes its existence to a celestial event.

Fort Venus, Tahiti, from *Journal of a Voyage to the South Seas,* Sydney Parkinson, 1784. Dixson Library, State Library of New South Wales

The consensus among astronomers was that the observations of the 1761 transit, made by a number of nations around the world, had not been wholly successful. Britain's Royal Society determined to rectify this in 1769, when the transit would take place on 3 June. The Society resolved in June 1766 to send observers to various parts of the world, including to an as yet unspecified location in the South Pacific. As will be seen below, astronomers needed observations from this area for comparison with those from other locations so that they could make meaningful calculations of the distance of the Sun.

To seek funds for this voyage the Society submitted a memorial 'To the Kings most excellent Majesty', mentioning that other European powers – France, Spain, Denmark, Sweden and Russia – were already making preparations to observe the 1769 transit. The memorial stressed that the first observation of a transit of Venus was by an Englishman, Jeremiah Horrocks, in 1639. It concluded with a patriotic appeal:

> That the British Nation has been justly celebrated in the learned world for their knowledge of Astronomy, in which they are inferior to no Nation upon Earth, Ancient or Modern; and it would cast dishonor [sic] upon them should they neglect to have correct Observations made of this important Phenomenon.

The Royal Society's request fell on fertile ground, for Britain had been keen to send an expedition to the South Seas to look for lands to expand its empire, but was afraid of offending the French, the Dutch and the Spanish, who all had interests in the area. The transit of Venus provided the cover of a scientific voyage for such a mission. In March 1768 King George III granted the Royal Society £4000 for the South Seas voyage, and the British government indicated its willingness to provide a Royal Navy ship with its crew.

A MASTER MARINER TAKES COMMAND

Alexander Dalrymple, a member of the British East India Company, was the first choice of the Royal Society to make the observations. He wanted full command of the ship, however, which the Admiralty would not allow because he was not a naval officer. The Admiralty then had to look for a suitable man to lead the expedition. This was no easy task. That man had to be a good leader, an excellent sailor, able to make astronomical observations, able to draw up maps of uncharted areas, and have good enough mathematical knowledge to understand the new method of finding longitude at sea promoted by the Astronomer Royal.

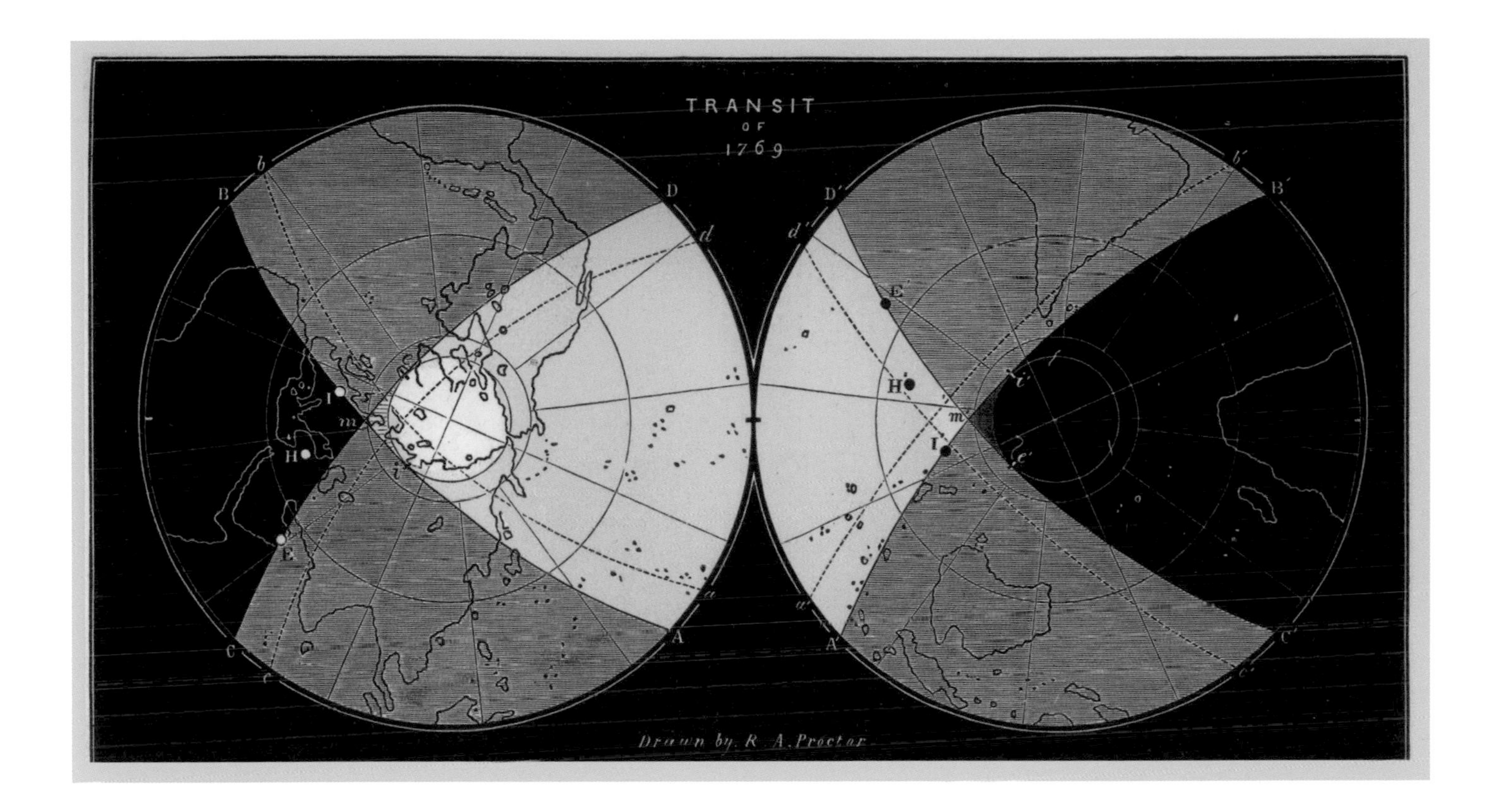

Fortunately, there was a person in the navy who met all the requirements, but he held the lowly post of ship's master and was not a commissioned officer. Since there was no one else suitable, the Admiralty had no alternative but to appoint James Cook a lieutenant and put him in charge. (Although Cook was officially only a lieutenant, any officer in command of a vessel was entitled to be addressed as captain.)

The next task was to find a ship. Cook recommended one of the ships used to carry coal, the type of vessel on which he had first learned seamanship. He indicated that these 'cat-built

The visibility of the 1769 transit of Venus, with the northern hemisphere on the left and the southern hemisphere on the right. The brightest area denotes the regions from where the transit was fully visible, the shaded areas indicate regions from where the beginning or end of the transit was visible, and the darkest areas are the regions from where the transit could not be seen. Map from *Transits of Venus* by RA Proctor, 1874. Brian Greig collection

At first apprenticed to a grocer, James Cook found his true calling – the sea.

An engraving of Captain James Cook published by JR Sherwin in 1784 based on the famous 1776 painting by Nathaniel Dance. Brian Greig collection

BORN IN THE YORKSHIRE village of Marton-in-Cleveland on 27 October 1728, James Cook was first apprenticed to a grocer, but soon moved to the seaside town of Whitby on the east coast of England. There he joined the crew of a ship in the North Sea coal trade. At the age of 26, although he was offered a command by his employer, he joined the Royal Navy as an able seaman. His ability must have been obvious because within a month he was serving as a master's mate. He gained his master's certificate two years later.

In 1758, after crossing the Atlantic as master of the ship *Pembroke*, Cook gained considerable experience as a surveyor in North America. He was largely self-taught, but he must have learned some of his surveying from a Dutch-born army captain, Stephen Holland, who was surveying the St Lawrence River at the same time as he was. Cook's surveys of Newfoundland, from 1764 to 1768, in particular, were highly regarded by the Admiralty. In August 1766, he had observed an eclipse of the Sun and his report of the event had brought him to the notice of the Royal Society as well. This prior recognition would have helped to minimise any opposition from the Royal Society to his appointment.

During his American survey, Cook returned briefly to England in 1762 and met and married a young woman named Elizabeth Batts, who at age 20 was 14 years his junior. There were to be six children of the marriage, of whom three lived to adulthood. The children rarely saw their father because he was almost constantly at sea. After the transit of Venus he sailed on two more great voyages of exploration. The last of these ended in tragedy when he was killed in Hawaii on 14 February 1779. Elizabeth was to survive her husband by over half a century, living to the age of 94.

[bluff-bow] barks' had good carrying capacity, could sail well on the open sea and had a shallow draught so that they could approach the shore closely. One of these, the *Earl of Pembroke*, was available and deemed suitable as it was of the right size and less than four years old. She was duly purchased for £2800. After a refit, she was awarded the grand name of the *Endeavour*. Cook joined the ship on 27 May 1768; it would be his home for nearly three years.

Earlier in the same month that Cook joined the *Endeavour*, the navigator Samuel Wallis had returned from a two-year voyage to the Pacific aboard the *Dolphin*. During this voyage he became the first European to see Tahiti and land there. He named his discovery King George's Island, and was able to provide its exact longitude and latitude. This was a stroke of luck, because Tahiti happened to be located almost in the centre of the area that the Astronomer Royal, Nevil Maskelyne, had suggested as the best position for a South Seas observation of the transit.

Maskelyne selected that area of the Pacific because observations of the transit of Venus from there would be of great use for both the method of durations advocated by Edmond Halley back in 1716 and the more recent technique of Joseph-Nicolas Delisle. The place where the duration of the transit would be the shortest on the globe was in that area, as well as the spots where Venus was to enter the Sun's disc (ingress) at the latest time and the spot where Venus was to leave the disc (egress) at the earliest time. Observations from Tahiti would thus be important for both methods of analysing the results.

Cook's instructions from the Admiralty were dated 30 July 1768. He was to sail to Tahiti to make his observations on 3 June 1769. In addition to this, his instructions read, 'When this Service is perform'd you are to put to Sea without Loss of Time, and carry into execution the Additional Instructions contained in the inclosed Sealed Packet'.

The sealed packet required Cook to sail south from Tahiti in an attempt to locate the fabled 'Unknown South Land' – *Terra australis incognita* – believed to lie in the southern ocean. If Cook failed in this, he was to chart the islands of New Zealand (the west coast of which had been charted by the Dutch mariner Abel Tasman in 1642) and return to England either via Cape Horn or the Cape of Good Hope.

These additional instructions were given extra relevance, if in fact they were not prompted, by Wallis's report that some of his men had seen what they believed to be mountain tops of the southern continent to the south of Tahiti. The Admiralty's instructions emphasised how the discovery of such a continent would be useful to trade and navigation.

Cook himself was to observe the transit, assisted by the 33-year-old astronomer Charles Green, formerly an assistant to the Astronomer Royal at Greenwich. In 1763 and 1764 Green, together with Nevil Maskelyne, had travelled out to Barbados in the West Indies to test John Harrison's fourth chronometer. Soon after their return to England from that voyage, Maskelyne had been appointed Astronomer Royal. Green either liked the shipboard life that he experienced or had a quarrel with Maskelyne, because he left Greenwich a few years later and joined the navy. Sadly, Green would be one of the casualties of the *Endeavour*'s voyage, dying of dysentery in January 1771 before reaching home.

Also on board the *Endeavour* – though not as part of the crew – was Joseph Banks. Born in 1743, Banks was young and rich; he was also passionate about natural history. He had been elected a Fellow of the Royal Society in 1766, not because of great scientific learning, but because of his social connections. Precisely how Banks negotiated passage on the *Endeavour* for himself and his party of eight is not known. The Royal Society formally recommended him to the Admiralty and their lordships agreed, no doubt influenced by the fact that Banks would pay for himself and his suite.

Banks brought with him two naturalists, Daniel Carl Solander and Herman Deidrich Sporing, both Swedes, to describe the plants, animals and birds that would be found. The artists Alexander Buchan and Sydney Parkinson were to draw their discoveries.

Banks' inclusion in the expedition would have many consequences. It was he who would, in 1779, recommend Botany Bay in New South Wales as a suitable place for a penal settlement.

At that time long sea voyages were dangerous affairs, mainly due to the disease scurvy, which we now know is mainly due to the deficiency of vitamin C. To try to control scurvy on the *Endeavour*, the ship carried among other provisions malt, which is grain that has been dried after sprouting, and sauerkraut (pickled cabbage). Cook also insisted on a strict regimen of daily washing for each crew member and the airing out of the decks. During the voyage there were to be no deaths from scurvy, a most unusual occurrence for the time. Cook attributed his success to the effect of the daily doses of malt that his crew had to take on his insistence. With the benefit of modern

Sketch of Matavai Bay, Tahiti, from the collection of Admiral Isaac Smith. Cook made his observations of the 1769 transit from the northeast point of the bay, which he named Point Venus. Mitchell Library, State Library of New South Wales

knowledge, the sauerkraut would seem more likely to have had an effect, although the processing of the cabbage would have destroyed most of its vitamin C content and hence its antiscorbutic property.

The *Endeavour* sailed from Plymouth on 25 August 1768 with 94 people on board. She called at Madeira in September, then sailed down to Rio de Janeiro, arriving in November, and in January 1769 reached Tierra del Fuego, the archipelago at the southern tip of South America. Crossing the immense waters of the Pacific, the *Endeavour* made landfall at Tahiti on 13 April 1769. No unknown land mass had been seen en route, leading Banks to comment that many square miles of the southern continent beloved of the speculative writers had been changed into water, and their ideas liquefied.

Cook was an excellent navigator and was one of the first sea captains to know the location of his ship at all times. Chronometers were still being developed at this time, and for guidance Cook had copies of the *Nautical Almanac* for the years 1768 and 1769. These volumes contained tables listing the angular positions of bright stars from the Moon at regular intervals. With these tables Cook could employ Maskelyne's method of lunar distances to determine the time at Greenwich and hence his longitude.

Cook anchored at Matavai Bay, as Wallis had done, and quickly decided on the northeast point of the bay, which he named Point Venus, as the place to make the observations. He also decided to erect a fort, appropriately named Fort Venus, for protection. Finished by 1 May, Fort Venus would accommodate about 45 men in tents, including the officers, the observatory, the armourer's forge and a cook's oven.

Cook described the fort in his log book:

> The North and south parts consisted of a Bank of earth 4½ feet [1.4 metres] high on the inside, and a Ditch without, 10 feet [3 metres] broad and 6 feet [1.8 metres] deep: on the west side faceing the Bay a Bank of earth 4 feet [1.2 metres] high and Pallisades upon that, but no ditch the works being at highwater mark: on the East side upon the Bank of the River was place'd a double row of casks: and as this was the weakest side the 2 four pounders were planted there, and the whole was defended besides these 2 guns with 6 Swivels and generally about 45 Men with small arms including the officers and gentlemen who resided aShore.

With the fort ready, Cook had the all-important quadrant, which was to be used to determine geographical location, taken ashore for the first time. To ensure the safety of this and the other instruments in the observatory, Cook placed a sentinel on watch all night. That precaution did not turn out to be enough, however. Somehow during the night an ingenious Tahitian managed to steal the quadrant as it lay in its case. Most likely he must have seen the great care with which the instrument was treated by Cook and his officers, and then deciding it was valuable, watched for an opportunity to grab it.

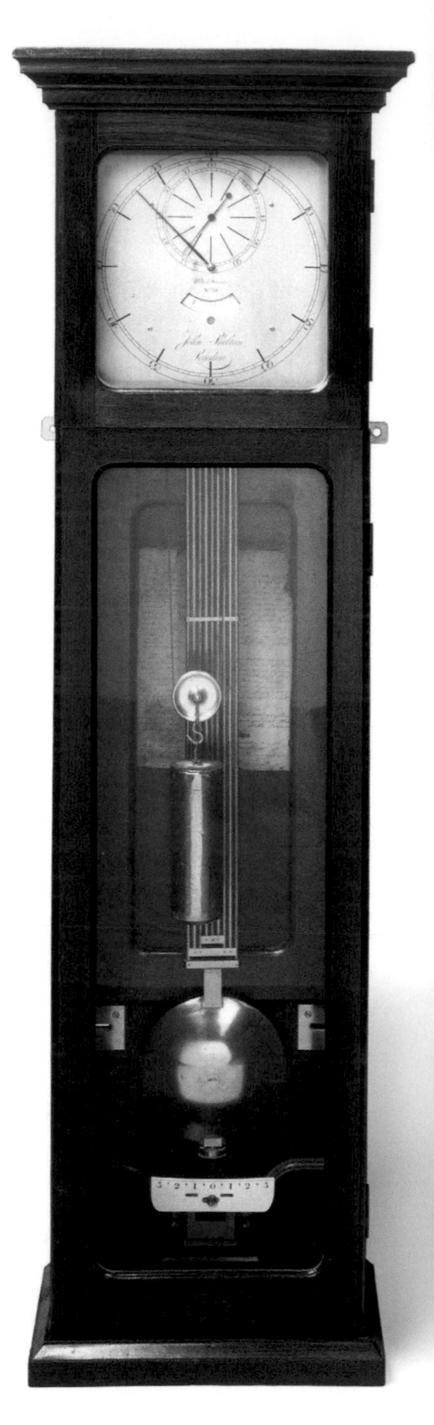

∧ The 12-inch (30-centimetre) astronomical quadrant made by the eminent London instrument maker John Bird and used by James Cook and Charles Green to determine the position of Fort Venus at Matavai Bay in Tahiti. The Royal Society London/ Richard Valencia

> One of five astronomical regulator clocks made by John Shelton that were used for transit of Venus observations in 1769. It is possible that this was the one Cook took to Tahiti. Much later, this clock (Royal Society 34) was taken to New Zealand for the 1882 transit. © Science Museum / Science & Society Picture Library

Cook was horrified; the theft put the success of the whole voyage in jeopardy. At this time Joseph Banks, who had maintained close and good relations with the locals, stepped in by questioning a friendly chief about the identity of the culprit. Soon Banks, accompanied by the astronomer Green, was in hot pursuit. As they pursued the thief, they were handed part of the quadrant and after considerable further enquiry the box with the remaining parts. Though Cook was pleased with the return of the instrument there was still a concern over any possible damage. Fortunately, Herman Sporing, one of the naturalists, had training as a watchmaker and could put the instrument together.

VENUS FROM TAHITI

On 3 June 1769, the day of the transit, Cook was fully prepared. Fort Venus was complete, with the clock made by John Shelton in one tent, and in the nearby observatory the quadrant made by John Bird, firmly supported on a large cask filled with wet sand. He and Green were to observe the transit from Fort Venus using identical Gregorian-type reflecting telescopes, from the workshop of the Scottish instrument-maker James Short, of 2-feet (60-centimetre) focal length. Observing with them was Banks' friend and fellow botanist, Dr Solander, who used his own telescope.

Cook described the preparations for the observations in his report to the Royal Society, which was published in 1771:

> The astronomical clock, made by Shelton and furnished with a gridiron pendulum, was set up in the middle of one end of a large tent, in a frame of wood made for the purpose at Greenwich, fixed firm and as low in the ground as the door of the clock-case would admit, and to prevent its being disturbed by any accident, another framing of wood was made round this, at the distance of one foot from it. The pendulum was adjusted to exactly the same length as it had been at Greenwich. Without the end of the tent facing the clock, and 12 feet [3.7 metres] from it, stood the observatory, in which were set up the journeyman clock and astronomical quadrant: this last, made by Mr. Bird, of one foot radius, stood upon the head of a large cask fixed firm in the ground, and well filled with wet heavy sand. A centinel was placed continually over the tent and observatory, with orders to suffer no one to enter either the one or the other, but those whose business it was. The telescopes made use of in the observations were – Two reflecting ones of two feet focus each, made by the late Mr. James Short, one of which was furnished with an object glass micrometer.

Matavai Bay and Point Venus. Tahiti.
Augt. 24th 1849.

Cook took precautions against overcast weather at Fort Venus by sending an observing party led by the *Endeavour*'s third lieutenant westwards to the island of Moorea and another party led by the second lieutenant to an island in the east. As it happened this was unnecessary, for the day was clear, though extremely hot.

Cook summed up the observations in his log book for 3 June:

> This day prov'd as favourable to our purpose as we could wish, not a Clowd was to be seen the whole day and the Air was perfectly clear, so that we had every advantage we could desire in Observing the whole of the passage of the Planet Venus over the Suns

< Sketch of Matavai Bay and Point Venus, Tahiti, in 1849, by Admiral Sir Edward Gennys Fanshawe. © National Maritime Museum, Greenwich, London

View of Point Venus and the *Endeavour* at anchor in Matavai Bay, Tahiti, from One Tree Hill with a tree of the *Erythrina* species in the foreground. This is an Italian version of the engraving in John Hawkesworth's official account of Cook's first voyage. Brian Greig collection

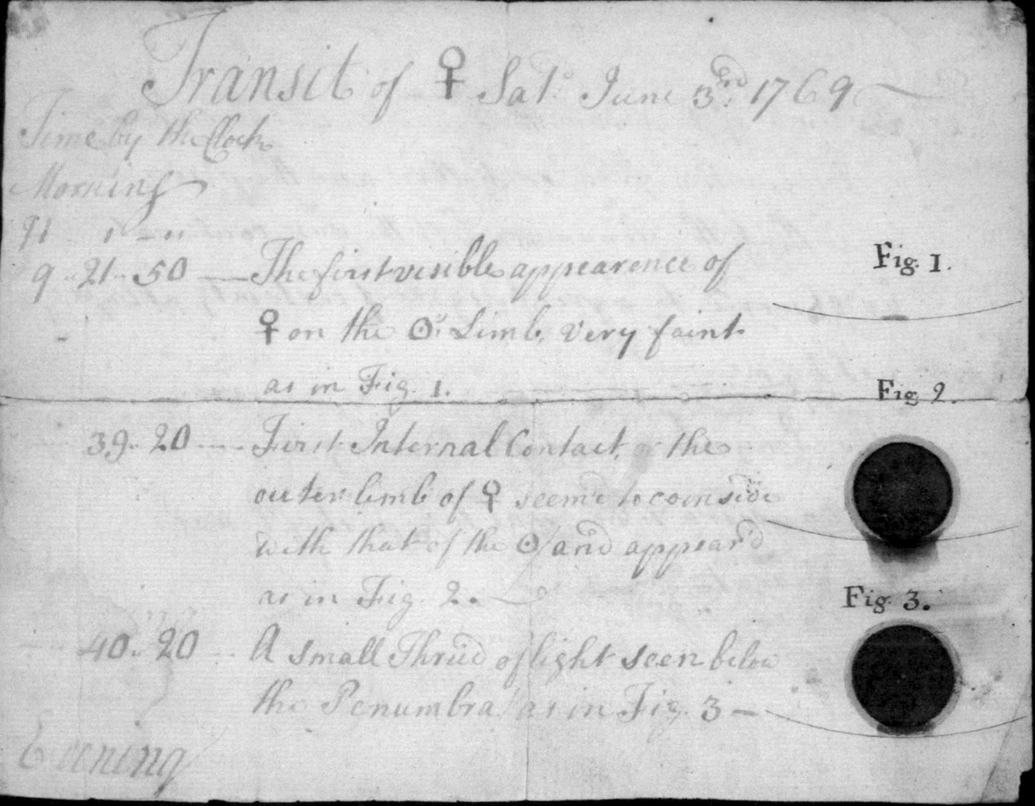

Transit of ♀ Sat. June 3rd 1769

Time by the Clock

Morning

H ′ ″

9 – 21 – 50 — The first visible appearence of ♀ on the ☉'s Limb, very faint as in Fig. 1.

Fig. 1.

Fig. 2.

39 – 20 — First Internal Contact, or the outer limb of ♀ seem'd to coinside with that of the ☉ and appear'd as in Fig. 2.

Fig. 3.

40 – 20 — A small Thread of light seen below the Penumbra as in Fig. 3 —

Evening

Cook's description of the transit. Mitchell Library, State Library of New South Wales, ZML S1/66

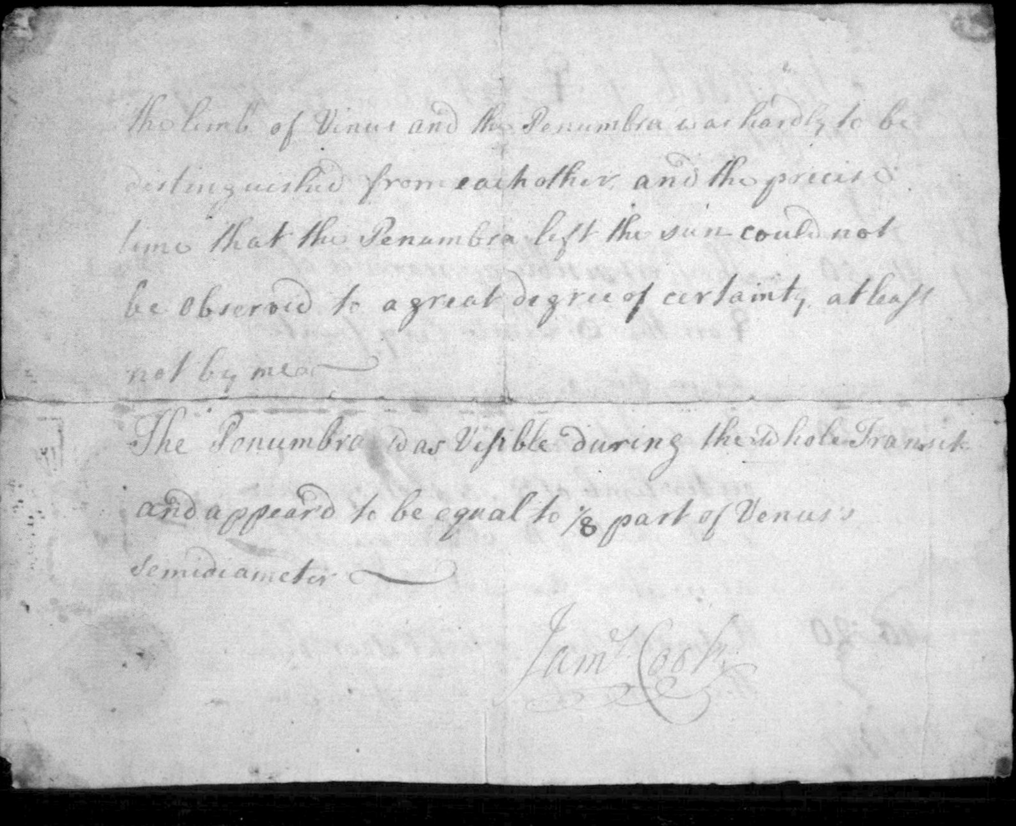

the limb of Venus and the Penumbra was hardly to be distinguished from each other, and the precise time that the Penumbra left the sun could not be observed to a great degree of certainty at least not by me —

The Penumbra was Visible during the whole Transit and appear'd to be equal to 1/8 part of Venus's semidiameter —

Jam.s Cook

Cook's description of the transit. Reverse side. Mitchell Library, State Library of New South Wales, ZML S1/66

> disk; we very distinctly saw an Atmosphere or dusky shade round the body of the Planet which very much disturbed the times of the Contacts particularly the two internal ones. Dr. Solander observed as well as Mr. Green and my self, and we differ'd from one another in observing the times of the Contacts much more than could be expected. Mr. Greens Telescope and mine were of the same Magnifying power but that of the Dr. was greater than ours. It was nearly calm the whole day and the Thermometer expose'd to the Sun about the middle of the Day rose to a degree of heat we have not before met with.

Both Cook and Green missed the exact time of the first contact of Venus with the Sun, their sightings being thrown off by an unexpected ring of light around the planet. They referred to this light as a 'penumbra', which presumably was similar to the 'halo' described by observers of the later 1874 transit. According to Cook, the width of the penumbra was one eighth of the semi-diameter of Venus, or about three seconds of arc. First contact was difficult to judge because the penumbra was dark near the planet's limb (edge). However, Cook said:

> At this time a faint light, much weaker than the rest of the penumbra, appeared to converge towards the point of contact, but did not quite reach it ... and was of great assistance to us in judging of the time of internal contacts of the dark body of Venus, with the Sun's limb.

In spite of the difficulties, Cook and Green agreed exactly on the time of the second contact, with Dr Solander recording a time 13 seconds later. Solander did not record a time for the third contact, but Cook and Green differed by only six seconds in its timing. Cook also found the last external contact difficult to time, as it was hard to distinguish between the edge of Venus and the penumbra. Although no one had expected to see the penumbra, Cook was confident that it was not an illusion, for not only had he and Green both seen it, but Dr Solander had seen it even more distinctly through his higher magnification telescope.

As we will see later, these results were excellent, even by the standards established at the following transit, over a century later. However, Cook and Green had had very high expectations about the accuracy to be expected, and were despondent at their results. So it was that despite these successful and well-planned observations, at this point Cook regarded his voyage to the South Seas as a failure – fortunately, this 'failure' was to be redeemed by his mapping of the coastline of New Zealand

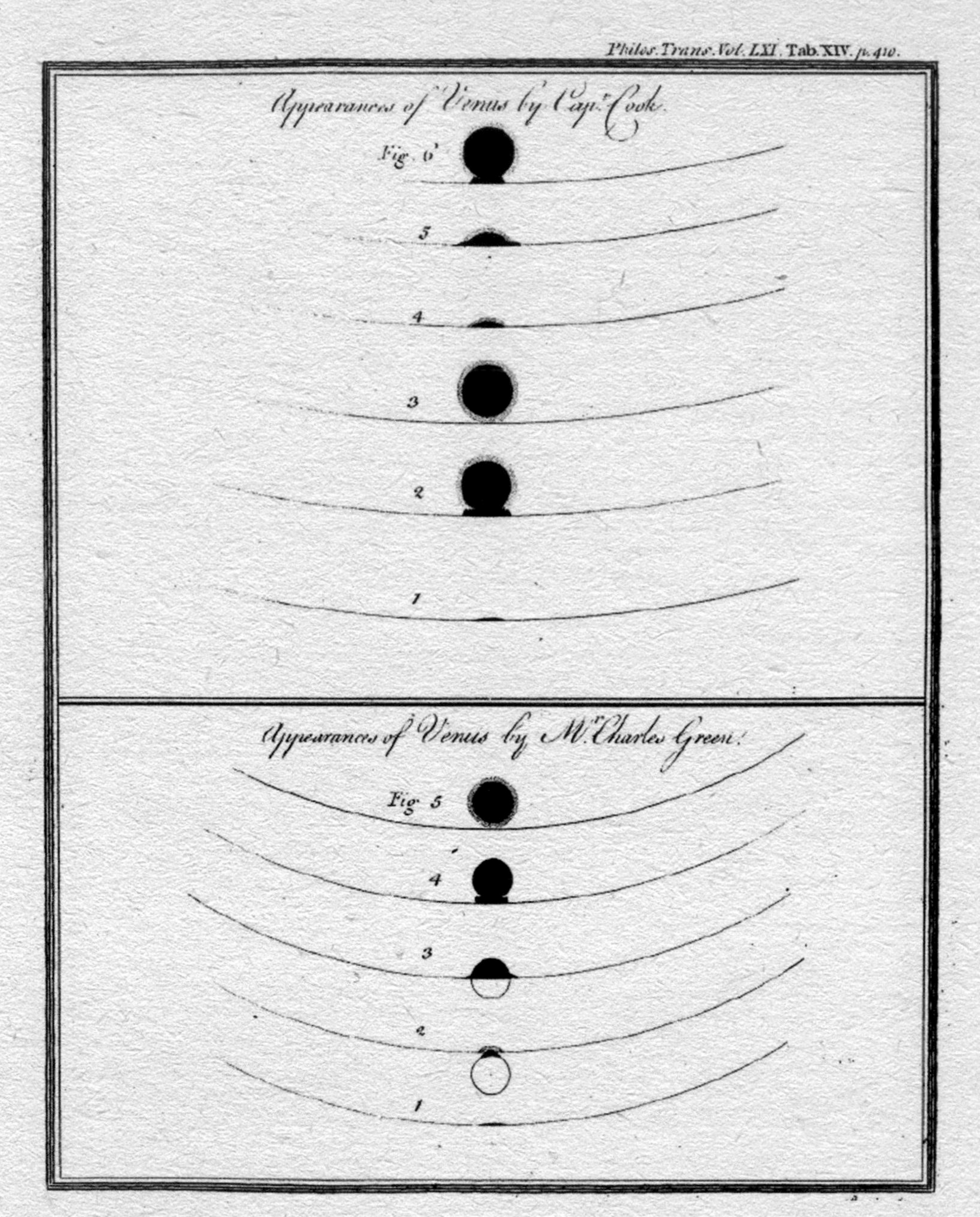

Cook's and Green's drawings of the 1769 transit. *Philosophical Transactions of the Royal Society*, 61: 410 (1771)

and the east coast of the land he called New South Wales.

Before and after the transit, Charles Green was busy at the observatory making astronomical observations. With Green's death on the return voyage, the Astronomer Royal, Nevil Maskelyne, made the necessary calculations and published the observations and results in the *Philosophical Transactions of the Royal Society* in 1771. One set of observations with the quadrant was to establish local time and the going rate of the Shelton clock, which was a loss of 20.9 seconds a day. Maskelyne was unimpressed with another set of observations with the quadrant to establish the latitude of Fort Venus because they fluctuated more than expected. He comments: '… the cause of which, if not owing to want of care and address in the observer, I don't know how to assign.' Of course, another cause for the fluctuations may have been damage to the quadrant when it was stolen, but Maskelyne may have been unaware of that episode.

The remaining astronomical observations from Tahiti were of the Moon so as to give longitude by Maskelyne's own technique of lunar distances. Green also observed the moons of Jupiter through the telescopes, again with the purpose of determining longitude.

With observations of the transit completed, Cook overhauled the *Endeavour* and charted the island of Tahiti, so it was not until 13 July that the expedition left on the continuation of the voyage. Cook did not immediately sail south but went first to examine the neighbouring islands, naming the group the Society Isles because of their proximity to each other.

On 9 August the *Endeavour* sailed south in conformity with Cook's instructions, in search of the Great South Land. On 2 September, in the face of heavy gales and with no sign of land, Cook turned northeast, then, three days later, northwest. Turning southwest on 21 September, he sighted the east coast of the North Island of New Zealand on 7 October. By circumnavigating both the North and South Islands, he proved that New Zealand was not part of any southern continent.

By March 1770, Cook had fulfilled his instructions from the Admiralty and could return home. But to sail the southern route via Cape Horn or the Cape of Good Hope would be dangerous during winter. He decided therefore to sail via the East Indies. *Endeavour* would leave New Zealand and sail 'westward untill we fall in with the East Coast of New Holland and then to follow the

No 20/50 "Botany Bay 'Welcome'"

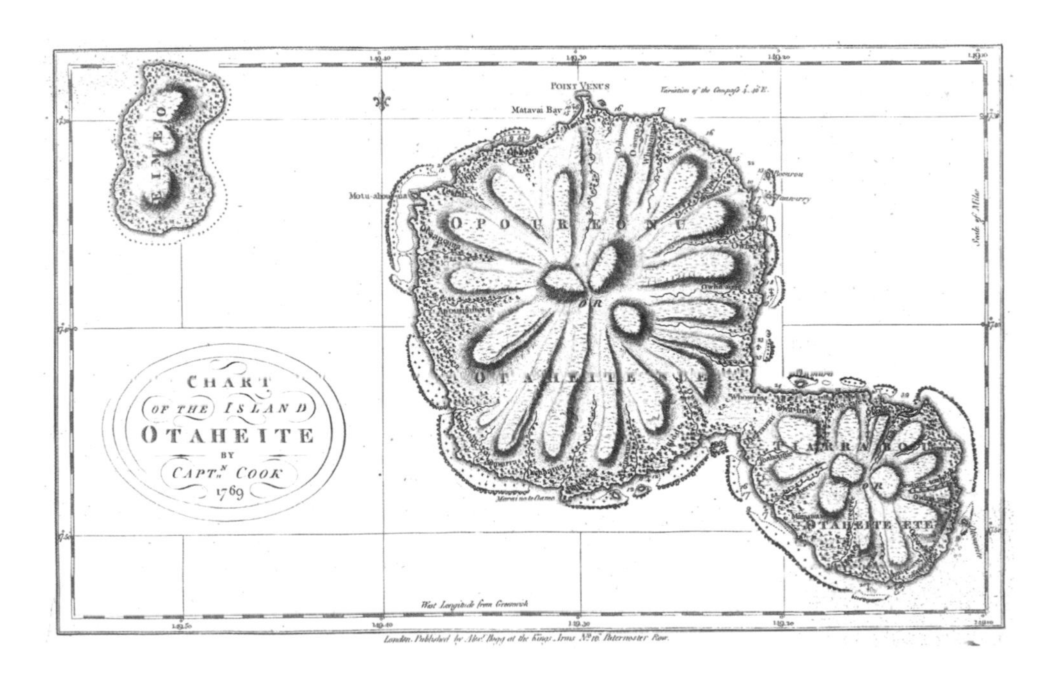

Cook's chart of Tahiti, published by Alexander Hogg of the Kings Arms, London. Brian Greig collection

H.M. Bark *Endeavour* in Botany Bay. Etching by Geoffrey C. Ingleton, 1937. National Library of Australia, nla.pic-an6152165

deriction [sic] of that Coast to the northward or what other direction it may take untill we arrive at its northern extremity'. That decision made Cook great.

The east coast of New Holland was a *tabula rasa* – a blank slate – which Cook charted, naming it first 'New Wales' and then 'New South Wales', and taking possession of it in the name of King George III at an island he named Possession Island off the tip of Cape York. He arrived back in England on 13 July 1771 a national hero.

Captain James Cook declares New South Wales a British possession at Botany Bay in 1770. Adapted from a painting presented to the Philosophical Institute of Victoria by the artist TA Gilfillan, but now lost. Brian Greig collection

A variety of optical effects disturbed their view, the most important of these being the black drop – a dark thread joining the edge of the dark silhouette of the planet to that of the Sun, making the instant of contact hard to determine.

Venus leaving the solar disc in visible light – the black drop effect captured by the TRACE satellite. NASA

EDMOND HALLEY HAD suggested that when observing the transit of Venus, the planet's ingress and egress on the Sun could be measured to within a second of time. Astronomers in 1761 and especially in 1769 discovered that was not possible because a variety of optical effects disturbed their view, the most important of these being the black drop. This is seen as a dark 'thread' or ligament joining the edge of the dark silhouette of the planet to that of the Sun and so making the instant of contact hard to determine. Many observers in 1761 and 1769 saw the black drop, including Maskelyne, who watched the start of the transit from Greenwich.

Cook and Green in their descriptions of the transit concentrated on the penumbra or haziness surrounding the planet. They did see the black drop, though, as is clear from their drawings, and from Cook's comment on the difficulties of timing the internal contacts 'by reason of the darkness of the penumbra at the sun's limb, it being nearly, if not quite, as dark as the planet'. If only the two of them had known that other observers had had to struggle with the same phenomenon, they would not have been so disappointed with their observations.

Astronomers have been trying to explain the cause of the black drop since the 18th century. It is only in recent years, after observations of a transit of Mercury and of the 2004 transit of Venus by spacecraft, together with numerical calculations, that scientists feel that they can explain the phenomenon. They say that the black drop takes place due to a combination of the normal blurring by the atmosphere, the effect of the size and quality of the telescope used for the observations and the well-known darkening of the disc of the Sun near its edge. Whether the black drop is seen and how clearly depends mainly on the conditions in the atmosphere and how far above the horizon the Sun is at the time it is observed.

A FATAL OBSESSION: THE FRENCH OBSERVERS

Cook and Green were possibly the most famous observers of the 1769 transit. However, among the many others who witnessed the event, two French astronomers stand out for the difficulties they endured and the diligence with which they carried out their tasks.

In the previous chapter we left Guillaume Le Gentil at Mauritius contemplating waiting for the 1769 transit instead of returning home to France. On deciding in favour of the transit, his first step was to sail on board a Spanish warship to Manila in the Philippines, arriving on 10 August 1766. Le Gentil had plans to make his observations from there, but he had difficulties with the Spanish governor, who accused him of forging his letters of recommendation. Having received a recommendation from France to go to the French enclave of Pondicherry in India, where he had originally meant to view the earlier transit, he left Manila on a Portuguese ship.

The voyage was fast, though not without incident. During a particularly perilous passage the captain and the pilot in charge of navigating the vessel had a loud disagreement, with the result that the pilot went below decks and shut himself into his cabin. With no one piloting the ship, Le Gentil had to take over his duties until the man was persuaded back on deck.

When Le Gentil arrived at Pondicherry on 27 March 1768, there was still over a year to go until the transit. There, with the help of the governor, he selected a suitable site for his observatory. It was on top of a ruined citadel and incorporated the remaining brick wall of a pavilion that had been part of a palace. Le Gentil seems to have been quite unperturbed by the knowledge that for some time 'sixty thousand weight of [explosive] powder' was stored under his observatory. Immediately the observatory was ready for occupation he moved his quadrant, clocks and other instruments there. By 14 July 1768 he could begin the task of making an accurate determination of the longitude and latitude of his location.

At last the day of the transit, 3 June 1769, arrived. Until the night before, Le Gentil was confident of a clear sky because the weather had been good throughout May and the first two days of June. That night, however, the wind woke him at 2 am and to his horror he saw that the sky was covered by cloud. By the time the clouds cleared away on the next morning the transit was over. Understandably, afterwards Le Gentil was in a state of shock.

A. *Attelier d'Artillerie.*
B *Magasin general.*
C. *Capucins.*
D. *Ruines des Capucins*
E *Maison.*
FF *Ruines de la Citadelle.*
G. *Mât de Pavillon*
HH. *Ruines du Gouvernement.*
HI. *Observatoire.*
K *Puis.*

VUE D'UNE

de la Gardette Sculp

A view of part of the ruins of Pondicherry in 1769 with the observatory built by the French transit expedition to the right of the flagpole. Plate 5 from Le Gentil's *Voyage dans les mers de l'Inde, fait par ordre du roi (A Voyage in the Indian Ocean, Made by Order of the King).* National Library of Australia

E DES RUINES DE PONDICHERY.
en 1769.

Le Gentil said:

> That is the fate which often awaits astronomers. I had gone more than ten thousand leagues; it seemed that I had crossed such a great expanse of seas, exiling myself from my native land, only to be the spectator of a fatal cloud which came to place itself before the sun at the precise moment of my observation, to carry off from me the fruits of my pains and fatigues …

Le Gentil was then anxious to return to France, especially as he had learned that his heirs were spreading the rumour of his death so that they could take over his estate. His bad luck, however, continued on his return journey. Due to illness he had once again to stop over in Mauritius, and when he did leave the island a storm so damaged the ship on which he was a passenger that it had to return there six weeks after departure.

After some unforseen difficulty in obtaining passage on a French ship, Le Gentil was offered a place on a Spanish warship and sailed on the ship to Cadiz in Spain. From there he travelled overland to France, crossing the border on 8 October 1771 after an absence of 11½ years. Even on his return he faced difficulties: his heirs were about to divide the estate from which a large sum had already been stolen and, most disappointingly, his place in the Academy of Sciences had been given to someone else.

Eventually, these problems were all sorted out and Le Gentil married and fathered a daughter. He lived happily for 21 years after his return until his death from illness in October 1792.

In the previous chapter, we saw how Abbé Chappe d'Auteroche travelled to Siberia to view the 1761 transit. For the 1769 transit, the French Academy of Sciences sent Chappe to the much warmer climes of Baja California, the peninsula on the western side of Mexico. He set out from Paris on 18 September 1768 together with a small entourage consisting of a servant and

> Mr. Pauly, the King's Engineer and Geographer, from whose talents I expected great assistance, was to second me in my astronomical and geographical operations: Mr. Noel, a pupil of the Academy of Painting, was intended for our draughtsman, to take draughts of sea coasts, plants, animals, and whatever we might meet with that was curious: lastly, Mr. Dubois, a watchmaker, was intrusted with the care of preserving my instruments, and repairing the little mischiefs they too often sustain in such long voyages.

En route at Cadiz in Spain Chappe was also asked by the Spanish Court to take two of their naval officers and astronomers, Messieurs Doz and Médina, with him to observe the transit. From Cadiz the party sailed to Veracruz in the Gulf of Mexico and then travelled overland across Mexico to the west coast. From there a small ship was to take them to Baja California.

The ship left on 19 April 1769, which was already close to the date of the transit, but was delayed by a nasty mixture of calm weather and contrary winds and currents. After a month at sea Chappe insisted on landing at the first possible place on the peninsula. This was a spot near San José del Cabo, a Spanish mission that had been founded four decades earlier. Though there were vigorous objections from the ship's pilot and the two Spanish naval officers on account of the strong winds and rocks at the spot, a successful landing was made using a number of trips by a small boat.

The whole party was accommodated in a large barn in which Chappe soon had the instruments set up and half the roof removed to allow observation. These instruments included two quadrants, one of 3 feet (0.9 metre) and the other of 1½ feet (0.45 metre) in radius; two telescopes by the English telescope-making firm of Dollond, one of 10-foot (3-metre) focal length and the other 3-foot; and a pendulum clock by the Swiss clockmaker Berthoud.

All would have been well except that there was a lethal epidemic spreading throughout the area that had already killed a third of the inhabitants of the local village. As there was so little time before the transit, Chappe resisted the advice of the Spanish naval officers to move further north and declared that he would stay to observe the transit whatever the consequences.

On 3 June 1769, with the sky clear, Chappe successfully observed both the ingress, and six hours later, the egress of Venus. He noted, though, that around the time of the second contact – that is, when Venus had just completely entered the disc of the Sun – the edge of the Sun appeared to attract the edge of the planet. He had described the black drop effect.

Two days after the transit most of the party became ill with the disease sweeping the area. At first Chappe was untouched and tended to the sick, but he soon succumbed as well. In spite of his serious illness he managed to observe an eclipse of the Moon on 18 June 1769 before dying six weeks later on 1 August.

For the surviving members of the party there was then a long wait for a ship to take them back to the west coast of Mexico and the journey home. Pauly, who may have been the only member of the French party to complete the return voyage, arrived back in Paris on 5 September 1770 with a casket containing Chappe's astronomical observations and notes from the journey. The Director of the Paris Observatory, César-François Cassini de Thury, later compiled these notes into the book *Voyage en Californie pour l'observation du passage de Venus sur le disque du soleil (Voyage to California to Observe the Passage of Venus across the Disc of the Sun).*

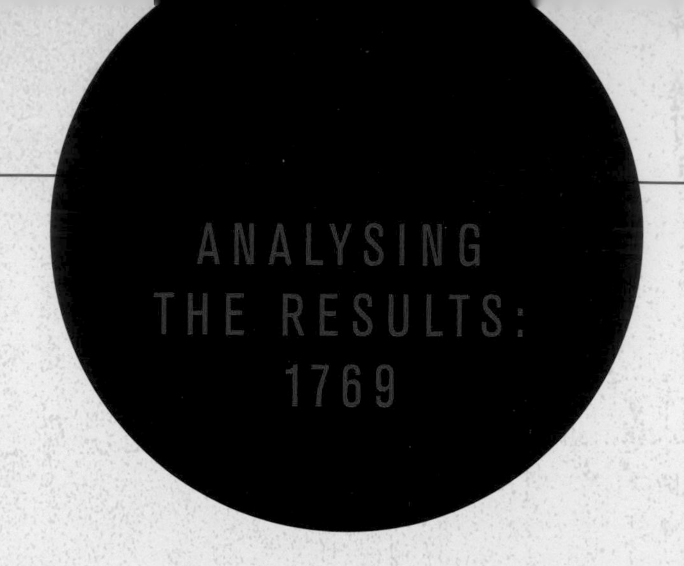

Many scientists attempted to calculate the distance of the Sun from the 1769 transit... up to 400 different calculations were presented to the Royal Society in London and similar learned societies in Europe.

MANY SCIENTISTS ATTEMPTED to calculate the distance of the Sun from the observations of the 1769 transit of Venus. In his 1874 book on the transit, the astronomy writer Richard Proctor estimated that a total of up to 400 different calculations were presented to the Royal Society in London and similar learned societies in Europe. Proctor listed five that he considered were the most carefully calculated, and these give values for the Sun's distance ranging from 154.8 to 148.1 million kilometres. Though there was still some uncertainty, the spread in the results was much smaller than the spread of almost 30 million kilometres in the 1761 results.

In 1824, many years after the two 18th-century transits, the German astronomer Johann Franz Encke published a new analysis of the observations of the 1769 transit in *Der Venusdurchgang von 1769 (The Passage of Venus in 1769).* In this he also combined the results with those of the earlier transit, obtaining a distance of 153.3 million kilometres. Encke's standing was such that for several decades astronomers all around Europe accepted this as the definitive value of the solar distance. Today, of course, we know that Encke's final value is too high.

Encke was one of the most eminent astronomers of his day. He was born on 23 September 1791 in Hamburg, Germany. After studying mathematics and astronomy and then fighting in a war against Napoleon's invading army, he was appointed to the observatory near the town of Gotha. There he began investigating the path of a faint comet discovered in 1818. In a series of articles in the following year he established that the comet had the same path as comets previously seen in 1786, 1795 and 1805, and so the three earlier discoveries were likely to be sightings of the same comet.

Encke predicted the return of this comet in 1822, with its closest approach to the Sun on 25 May. On 2 June 1822 John Dunlop and Carl Rümker at the Parramatta Observatory, near Sydney, Australia, recovered the comet using Encke's predictions. Halley's Comet had been the first to have its return predicted and then confirmed. The one found at Parramatta became the second. Its recovery not only made Encke famous, but also put the new observatory in the faraway land of Australia on the world scientific map.

Johann Franz Encke. The German astronomer reanalysed observations of the 1769 transit in1824 to make a new calculation of the Sun's distance from the Earth. Dibner Library of the History of Science and Technology, Smithsonian Institution Libraries, Washington, DC

In honour of Encke's work the comet was named after him.

In 1822 Encke became the director of the observatory in Gotha where he had been working. Three years later the king of Prussia instructed him to set up a new observatory in the capital city, Berlin. There he discovered a break in the rings of the planet Saturn that also added to his standing and which scientists still call the Encke gap.

This small brass telescope, made by Banks of London, was used at Parramatta Observatory in 1822 to recover Encke's comet. Collection: Powerhouse Museum, Sydney. Photo: Sue Stafford

CAPTURING THE TRANSIT

1874 AND 1882

The United States Transit of Venus Commission decided early on that, unlike the British observing teams, which were planning to put most of their efforts into timing the contacts at ingress and egress visually, the American teams would concentrate on making their observations photographically. The Commission thought the new technology would provide a greater number of results that were less subject to the vagaries of individual observers, and so the final result for the distance of the Sun would be more accurate.

Participants in the eight American transit of Venus expeditions of 1874 gathered at the old US Naval Observatory in Washington. Photo: The James Melville Gilliss Library, US Naval Observatory

NEW TECHNOLOGIES BRING A CLEARER VIEW

More people observed the 1874 transit than any previous transit, a reflection of the great expansion of technology, science, education, industry, trade and population over the preceding decades.

Many of the events that took place from the middle of the 19th century until the two transits of Venus of that era have helped to shape the world with which we are now familiar. For instance, in 1853 the American Commodore Matthew Perry sailed into Tokyo harbour with a squadron of four warships and demanded that Japan end its isolation from the West and enter into a trade treaty with the United States. The Japanese had no alternative but to agree.

A number of wars in Europe, culminating in the Franco-Prussian War of 1870–71, led to the unification of the German-speaking states into one nation under the leadership of a Kaiser. Similarly the various states on the Italian peninsula joined into the Kingdom of Italy at this time. The American Civil War of 1861–65 temporarily broke apart the United States; even after the northern victory, it would take decades to mend the divisions between North and South.

The Suez Canal opened in 1869, joining the Mediterranean Sea with the Red Sea and allowing ships to sail from Europe to Asia without having to waste time navigating around Africa. Along the same principle of speeding up trade and communication, the first successful telegraph cable between continents began operating between Europe and America in 1866. The telegraph was to play an important role in the observations of the 1874 and 1882 transits of Venus.

In 1874, the year of the first transit of the 19th century, a group of artists whose works were rejected elsewhere put on their own exhibition in Paris. A critic writing a very negative review titled his piece, 'The exhibition of the impressionists', and unwittingly gave a name to a new and very popular movement in art. In music the Russian composer Pyotr Ilyich Tchaikovsky's much-loved *Piano Concerto No. 1* premiered in 1875, while Italian composer Giuseppe Verdi's grand opera *Aida* had its first performance in 1871. Two classics of literature, Leo Tolstoy's *War and Peace* and *Anna Karenina*, were published in 1869 and 1877.

Technological breakthroughs directly affected astronomy. The most important technique available to present-day astronomers is spectroscopy, the ability to break light from distant objects into its component colours. Astronomers can analyse the resulting spectrum to deduce velocity, temperature, chemical composition and many other details of the celestial

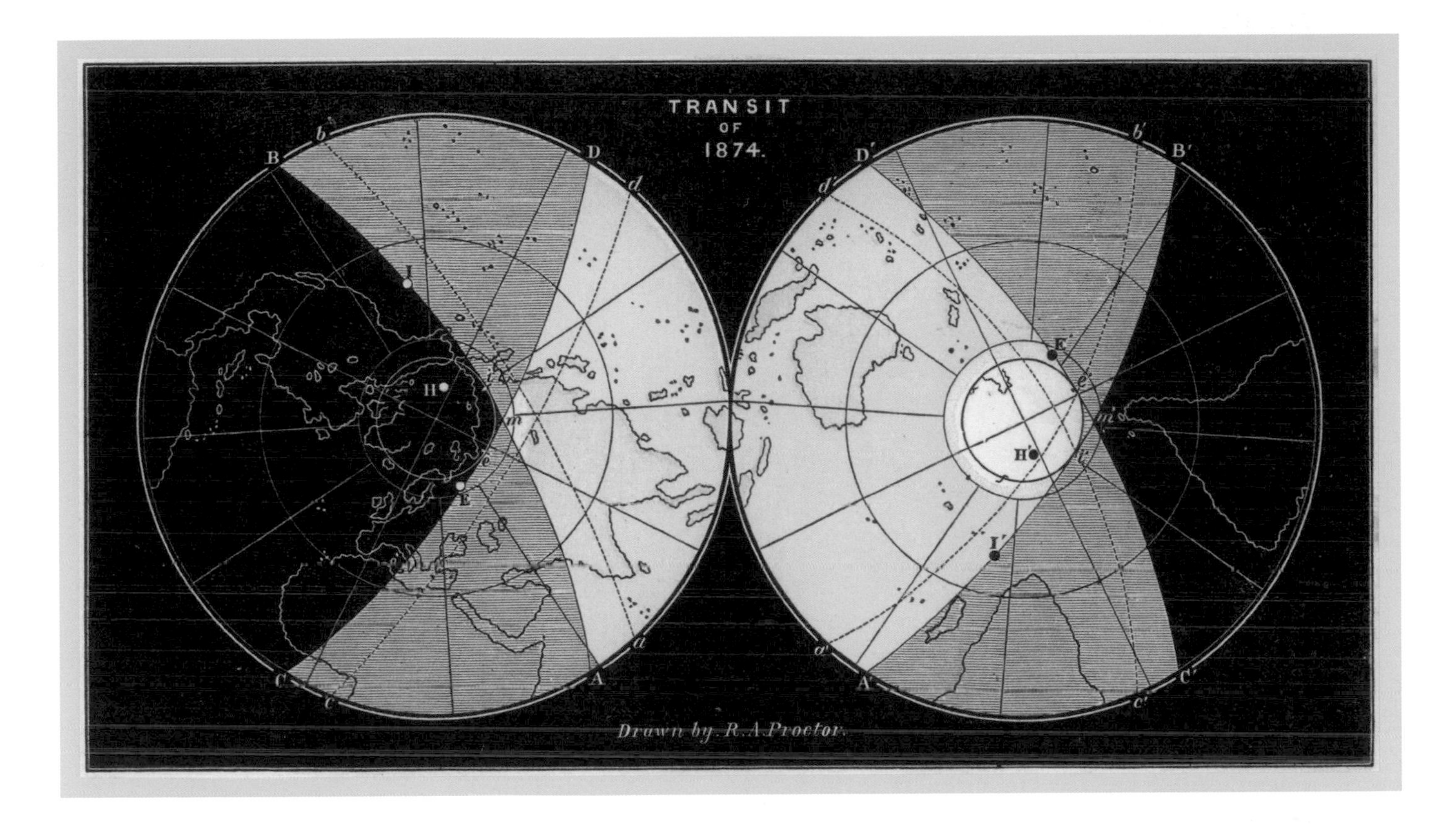

The visibility of the 1874 transit of Venus, with the northern hemisphere on the left and the southern hemisphere on the right. The brightest area denotes the regions from which the transit was fully visible, the shaded areas indicate regions from which the beginning or end of the transit was visible, and the darkest areas are the regions from which the transit could not be seen. Map from *Transits of Venus* by RA Proctor, 1874. Brian Greig collection

objects under observation. The 1860s saw the first use of spectroscopy for astronomy. A pioneering practitioner of the technique was a British amateur astronomer, William Huggins, who established that some nebulae (blurry objects in the sky) are only made of gas and do not contain any stars.

Photography was another important new technique in astronomy. American scientist John William Draper made the first photograph of the Moon in 1840, and ten years later another American astronomer, William Cranch Bond, managed to record an image of the star Vega.

Benefits of the technique were two-fold: astronomers could objectively record their observations and, because the photographic image builds up over time, they could 'see' features on the photographs that are invisible to the eye. Over the rest of the century there were steady improvements in the technique to make it more convenient and practicable as well as making the photographic plates more sensitive to light. Photography was to assume a key role in many of the transit of Venus observations in 1874 and 1882.

A feature of the 1874 transit was the observational efforts of 'new' states such as the United States and Australia. Since there was no visibility in North or South America, America sent eight international expeditions. Three American groups went to northern hemisphere sites – Nagasaki in Japan, Vladivostok in Siberia, and Peking (Beijing) in China. The other five groups went to southern hemisphere sites: Kerguelen Island in the South Indian Ocean, Chatham Island in the Pacific Ocean, Queenstown, New Zealand, and Hobart and Campbell Town on the island of Tasmania, Australia. Australia was the perfect place for observations, as the transit was visible from beginning to end, and each of the self-governing British colonies in Australia established observatories, while local observers established themselves in New South Wales, Victoria and South Australia.

Most European nations sent teams of observers to various places to make observations for the 1874 transit. France, for example, sent six parties, comprising altogether about 50 people, to Campbell Island, an uninhabited sub-Antarctic island to the south of New Zealand, to Saint-Paul Island in the Indian Ocean, to Noumea, the capital city of the French territory of New Caledonia, to Peking (Beijing) in China, to Saigon (now Ho Chi Minh City, Vietnam) in French Indochina, and to Yokohama in Japan. Britain sent observing teams for the transit to the Sandwich Islands (now Hawaii), to Egypt, to Rodrigues Island, which is in the Indian Ocean to the northeast of Mauritius, to New Zealand and to Kerguelen Island, in the southern Indian Ocean. In most of these places, the British set up subsidiary stations to increase the likelihood of at least one making successful observations.

> American sailing ship used to transport team to view transit of Venus in 1874. Thought to be in Vladivostok, Russia. Photo: The James Melville Gilliss Library, US Naval Observatory

AMERICANS IN THE NORTHERN HEMISPHERE

Each American expedition was equipped with a complex set of instruments to take the photographs planned by the Transit of Venus Commission. Some of the photos were to provide better timing of the contacts, while others would show the track of Venus across the Sun.

To make the images of the Sun as large as possible so as to allow accurate measurement, the commission chose a telescope of 40-foot (12-metre) focal length, which gave a 4-inch (10-centimetre) size image. As operating such a long telescope in the normal fashion would be awkward, the telescope was made stationary and horizontal. Sunlight was to be reflected into the far end of the telescope by a moving mirror driven by clockwork, a system called a heliostat. The mirror was an unsilvered wedge-shaped glass with the reflection of the Sun only from the front face reaching the telescope.

At the focus end of the telescope was a plate-holder for the photographic plates. In front of the plate-holder were a plumb line and a grid of ruled lines, which were both to be recorded on each photographic plate to assist with its subsequent measurement. Hollow iron piers supported the heliostat and the plate-holder, and a hut known as the photographic house enclosed the plate-holder. There were two other houses. One for a telescope with a 5-inch (13-centimetre) lens made by the famous American telescope manufacturer Alvan Clark & Sons. The other for a transit telescope, designed by William Harkness with an unusual broken-tube shape and used for determining time, longitude and latitude.

Each American expedition had the same instruments and carried with them the metal piers as well as the three prefabricated huts or houses. Each also had basically the same structure of personnel: one chief of party and chief astronomer, one assistant astronomer, one chief photographer and two assistant photographers. The people chosen as chief photographers were all photographic professionals, while their assistants were 'young gentlemen of education'. Before departure all the teams gathered in the grounds of the US Naval Observatory at Foggy Bottom, Washington, DC. There they practised observing the transit – with the help of a model of the planet passing in front of the Sun – as well as the steps involved in taking and processing the photographs.

Stackpole # 1507 broken-tube transit telescope designed by William Harkness and used by US Coast Survey staff on the transit of Venus expedition 1874 at Nagasaki, Japan. Photo: The James Melville Gilliss Library, US Naval Observatory

Five-inch (13-cm) equatorial Clark telescope in position as part of the 1874 American expedition in Nagasaki, Japan. Photo: The James Melville Gilliss Library, US Naval Observatory

AMERICANS IN NAGASAKI

One of the three northern hemisphere observing teams was sent to Japan. It was led by English-born George Davidson of the US Coast Survey. Due to his extensive surveying work, there are a number of places named after him including Mount Davidson, the highest natural point in the city of San Francisco, California. The two assistant astronomers on the team were also from the Survey.

Davidson, accompanied by his wife and son, and together with the rest of the team left San Francisco on 29 August 1874, which was five weeks after the other two northern hemisphere parties. They arrived at the port city of Yokohama on 23 September and immediately sent the instruments off to the selected city of Nagasaki. Davidson stayed in Yokohama for a short while to make arrangements with the Japanese government for the team's visit so that when he too left for Nagasaki his team was enlarged with additional members, including an interpreter and a couple of young officers from the Imperial Japanese Navy.

The city of Nagasaki is the capital of the Japanese island of Kyushu. On 9 August 1945, Nagasaki was the site of the second of two atomic bombs the United States dropped on Japanese cities near the end of World War II. However, in 1874 nuclear weapons had not yet been invented.

On arrival Davidson selected an observing site on a steep hill about one mile (1.6 kilometres) from the city centre. As there was no road access, his first step was to arrange for a road to be built so that the prefabricated houses, instrument piers and instruments could be carried up and installed.

Davidson reports that the morning of the transit began badly with not just one, but two layers of cloud forming:

> an upper stratum of cirrus and cirro-stratus forming a tolerably coherent screen, whilst below that and resting on a mountain 3 or 4 miles to the south of us (and 1,950 feet in elevation) was a dense stream of cumulo-stratus coming up slowly from the southwest with a steady movement.

There were breaks in the clouds, however, allowing a slightly late observation of first contact as Venus moved onto the Sun and a good observation of the second contact. Observation of the third contact was again slightly late due to inopportune clouds and it was not possible to observe fourth contact. Despite the clouds 116 photographic plates were taken. Davidson considered half of these as good images. Unfortunately, none of these appear to have survived.

AMERICANS IN PEKING

Canadian–born James Craig Watson, director of the observatory of the University of Michigan, was the chief of the American transit party sent to China. From the party's starting point in Chicago, the trip to the destination was a complex one involving many stages. First stop was San Francisco, from where they departed on the steamship *Alaska* for Yokohama, Japan, on 28 July 1874.

After transferring to another ship, they continued to Nagasaki where a US Navy ship was waiting to take them to the city of Tien–Tsin (Tianjin) in China. There Watson made enquiries about weather prospects at various localities and decided upon the capital city of Peking (Beijing) as the most likely spot for a clear sky on the day of the transit.

Once the instruments and their enclosures had been set up, the preliminary observations began. Watson made an important discovery on the night of 10 October when he found an asteroid, one of the many rocky bodies circling the Sun between the paths of Mars and Jupiter. As this was the first such object to be discovered from China it created considerable local interest with several 'mandarins of high–rank' visiting the observing station to look at it through the telescope. With great diplomatic tact, Watson invited Prince Kung, the regent of the Chinese Empire, to name the new asteroid. It is today still known as 139 Juewa, meaning 'the star of China's fortune'.

The day of the transit went well from a scientific point of view. Though there were some clouds, all four contacts were successfully observed visually and a good number of photographs were taken.

The Chinese authorities kindly assisted observations by closing the nearby street to carts, so as to stop interference from their vibrations. However, due to an unfortunate coincidence the American observers were in some danger in the period immediately after the transit. Watson writes:

> and it so happened that before the close of the transit a small-pox pustule came out upon the Emperor's face, so that there was a general belief that the prediction I had published was not after all a Transit of Venus, but that the Sun in the heavens by this spot exhibited an affliction which his representative on Earth, the celestial Emperor of China, was to suffer.

Smallpox is now eradicated, but then it was a serious disease that killed about a third of those who contracted it. As was made clear to the Americans, were the Emperor to die they would be

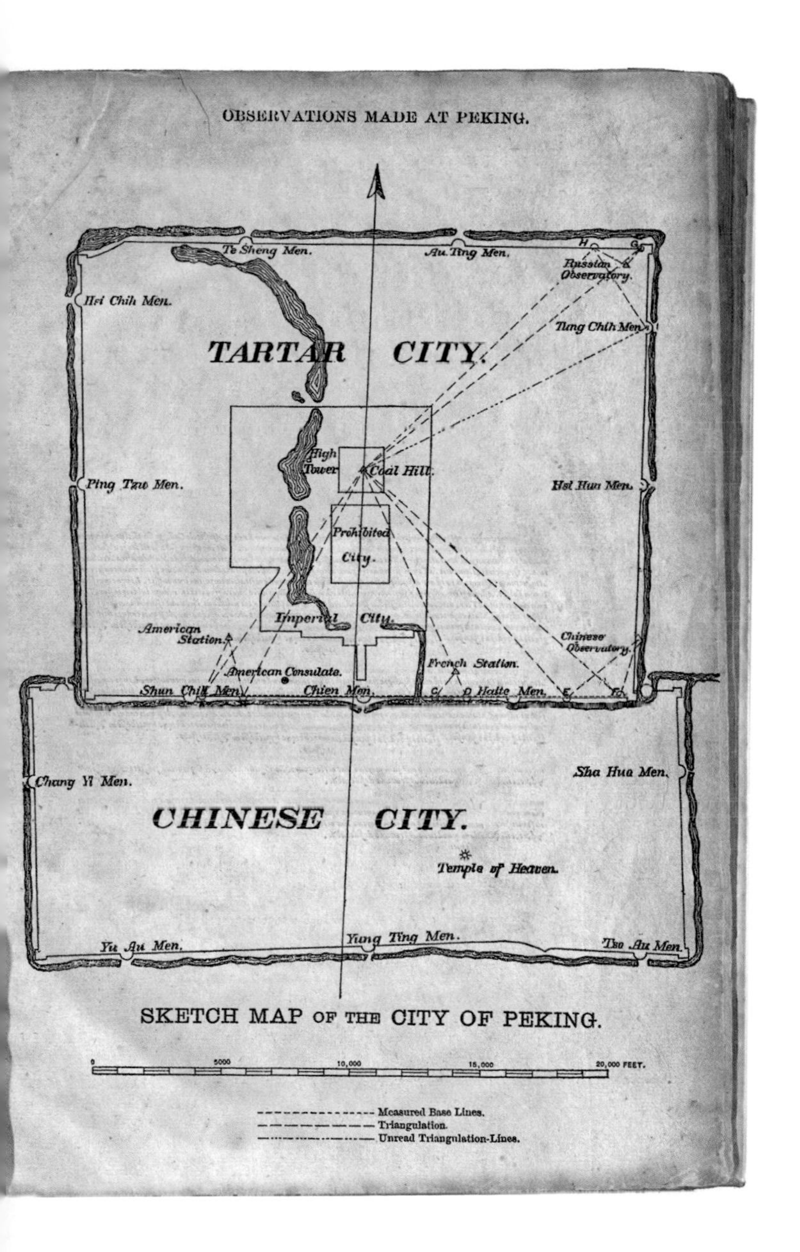

in grave danger, so they made enquiries about possible routes for a quick get-away from Peking. Fortunately, the Emperor appeared to recover and the team was able to depart in an orderly fashion. By the time the Emperor took a turn for the worse and died they were safely out of the city and on their way home.

Sketch map of Peking (Beijing) indicating the location of the American observing station. The dashed lines indicate some of the sight lines for surveying that the astronomers from the US transit of Venus expedition observed from the top of Coal Hill or Jingshan Park, the 150–foot high artificial mound just to the north of the palace. Source: Observations of the Transit of Venus, December 8–9, 1874, made and reduced under the direction of the Commission created by Congress. Part II. Section III. Observations made at Peking, under the direction of James C Watson

THE AMERICANS IN THE SOUTHERN HEMISPHERE

The remaining five American teams went to the southern hemisphere and departed aboard the USS *Swatara* a couple of months before those that went to the Northern Hemisphere. The *Swatara,* captained by Commander Ralph Chandler, left New York harbour on the early morning of Monday, 8 June 1874. By August, the *Swatara* was on course for the Crozet Islands, an uninhabited group administered by the French, halfway between Madagascar and Antarctica. This was where the team led by Captain Charles Walker Raymond meant to base themselves, but bad weather prevented a landing and, on 2 September, Captain Chandler decided to proceed to the other destinations.

The equatorial house containing the 5-inch (13 centimetre) lens telescope by Alvan Clark, set up in Barrack Square, Hobart, by one of the two American expeditions sent to Tasmania for the 1874 transit. Photo: William Harkness. George Eastman House. International Museum of Photography and Film

Next stop was the French possession of Kerguelen Island, to the east of the Crozet Islands, where an American team disembarked, joining a British group of observers. From there the *Swatara* proceeded toward Australia. At 4 pm on the afternoon of Thursday, 1 October 1874, the *Swatara* anchored in the harbour of Hobart Town, on the island of Tasmania, in the southernmost reaches of Australia, where two separate American expeditions disembarked, one led by Professor William Harkness and the other by Captain Charles Walker Raymond. Raymond's instructions were now to land at Hobart with a view to setting up an observation station in the 'vicinity of Melbourne'.

Harkness selected a site at Barrack Square, the location of the Anglesea Barracks, which had had a variety of civil uses since the last British regiment left Tasmania in 1870. Harkness records that at the time of his visit most of the buildings were 'rented to private families'. The elevated site was well suited to observing.

The party's first task was to install the piers for the instruments and then to put up the three observing huts so that the instruments could be set up inside them. These were then adjusted and tested. All was ready for the big day.

On the morning of 9 December, 'the heavens were black and lowering'. Over the next few hours there were occasional signs of improvement, but they all proved to be illusory. Eventually, about three hours after the transit had begun, the clouds cleared sufficiently to allow the team to take the first picture. Altogether, Harkness' party managed to take 41 images of the whole Sun. Then as internal contact at egress approached, they took a further 74 smaller images.

Harkness went to the equatorial house containing the Alvan Clark lens telescope to observe the third contact visually. He managed to make observations just before contact and then just after, with clouds blocking the view in between. He timed the instant after the contact carefully, though he could not be certain exactly how much time had elapsed since the contact.

The transit house containing the broken-tube transit telescope for obtaining time and longitude, in Barrack Square, Hobart. The picture shows members of the American team with the New Zealand War memorial in the background. Photo: William Harkness. George Eastman House. International Museum of Photography and Film

An essential task was to find the longitude of the observing site. The best way to do this was by telegraph from Melbourne. Comparison of time signals between clocks at Melbourne Observatory and at Hobart allowed the time difference between the two places to be found. As this time difference is equivalent to the two locations' longitude difference, the longitude of Barrack Square, Hobart, was easily obtained by adding the already known longitude of Melbourne. This was the first time that such a telegraphic exchange of clock signals had been attempted using a submarine cable in Australia.

His observations completed, Harkness left Australia from Sydney in March 1875 on board the steamship *Mikado* and returned to the Naval Observatory in June. There he spent the next two and a half years 'reducing' (that is, analysing) the observations from Hobart and the seven other US observing stations.

Although Captain Raymond's instructions were to set up an observatory in or near Melbourne, when

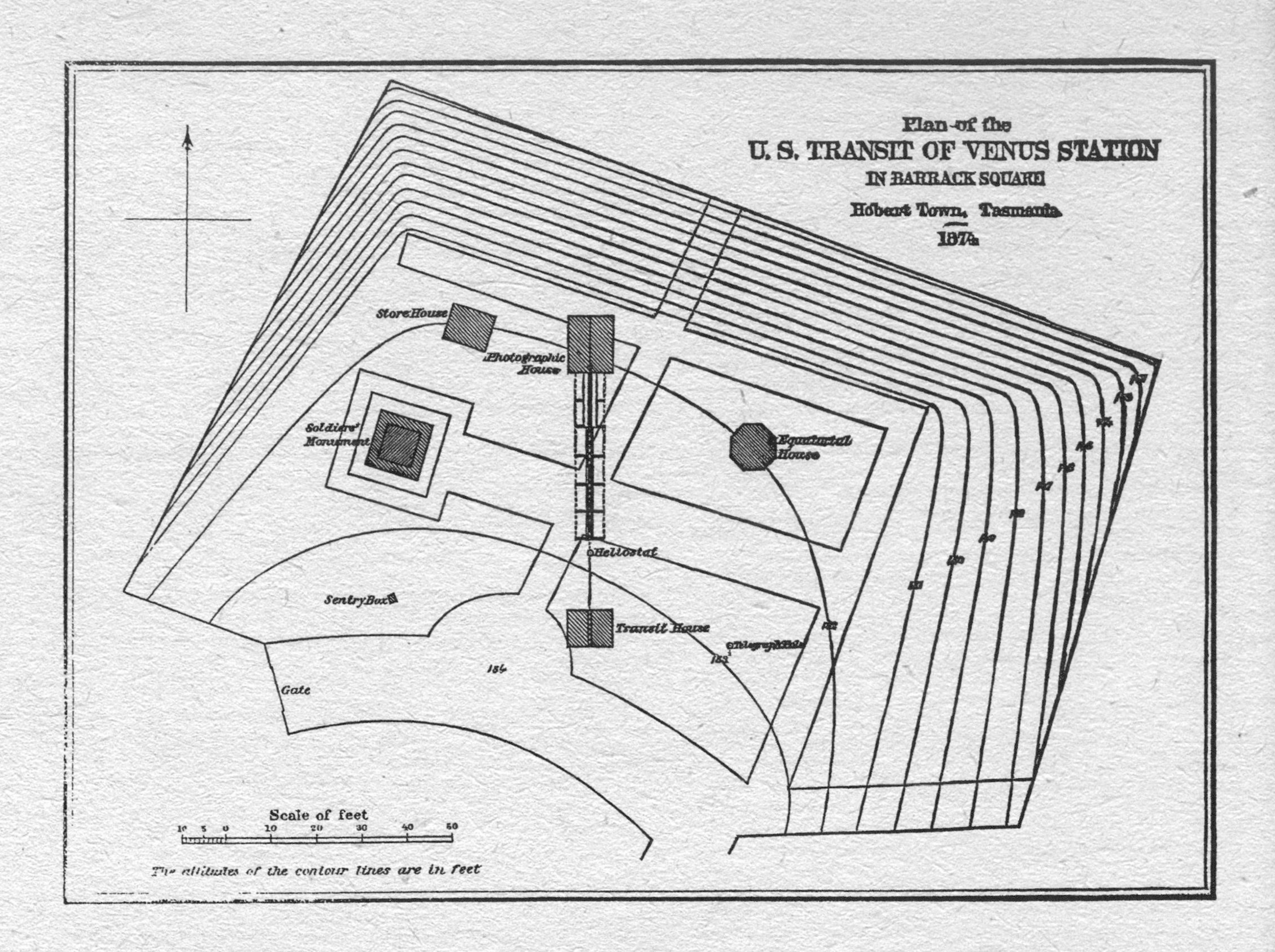

Plan of the US transit of Venus observing station in Barrack Square, Hobart, with contour lines indicating elevation. *Observations of the Transit of Venus, 8–9 December, 1874, made and reduced under the direction of the Commission created by Congress.* Part II. Section V. Observations made at Hobart Town, under the direction of William Harkness

William Harkness began his career as a doctor but, after service in the American Civil War, left medicine to join the US Naval Observatory, in which capacity he led one of the American teams observing the 1874 transit from Tasmania. The James Melville Gilliss Library, US Naval Observatory

he reached Hobart, a Dr William Valentine offered to host the party at The Grange, his property in Campbell Town, about 64 kilometres to the south of Launceston in the northern part of Tasmania. On looking at the available weather records, Raymond found that Campbell Town was justifiably regarded by the locals as the driest place on the island and for the previous nine years had had clear weather on 9 December. The main American party set off for Campbell Town on 9 October, with Raymond and assistant astronomer Lieutenant Tillman leaving a day later by mail coach carrying the two all-important chronometers 'in a carefully stuffed basket'.

By 10 November, the photographic house was completed and all its equipment installed. From then on until the transit, Raymond made his party practise the necessary procedures by taking test plates – about 500 of them. The team also practised using the Alvan Clark lens telescope and its double image micrometer. Raymond had a large white board painted with the different phases of the transit set up on a house about 1.5 kilometres away. Observing this board from the equatorial house gave the team plenty of practice in measuring the diameter of Venus and other features.

ANALYSING THE AMERICAN RESULTS: 1874

ON HIS RETURN TO THE US NAVAL Observatory in June 1875, William Harkness found himself in charge of measuring the plates returned by all eight American transit expeditions. He soon found that the plates taken as Venus was moving onto or off the Sun (that is, those taken between first and second contacts and between third and fourth contacts), had the same problem with the black drop and other optical effects as visual observations. Hence these could not be used. However, 221 images taken when Venus was fully on the Sun (that is, between second and third contacts) were giving good results.

A lack of funds meant that measurement and the necessary calculations did not progress easily. On three separate occasions between 1876 and 1882, the computers – people doing the laborious calculations – had to be discharged because there was not enough money to keep employing them. Situations like this, in which governments allocate funds for high-profile purposes such as a scientific expedition or a new hospital, but not for on-going maintenance, occur even in the present day. In the case of the transit of Venus, the attitude of Congress is at least partially understandable because the analysis took longer than expected and was showing limited promise.

There were similar problems with publishing the observations and results. Four volumes were projected and only one was published. Part II, which contains the reports of the eight observing parties, only reached the proof stage and only one printed copy exists. The last two volumes, which were meant to contain the results, were never published.

This vacuum without official results was pierced by a young assistant in the US Nautical Almanac Office, David Peck Todd, who published a preliminary result in 1881 based on 213 measured photographs. Of those photographs, 45 came from Nagasaki (Japan), 26 from Peking (China), 37 from Hobart (Tasmania, Australia), 32 from Campbell Town (Tasmania, Australia) and 45 from Queenstown (New Zealand) with only small contributions from the other three observing stations. Thus in spite of the difficulties with the weather at most of these sites, a useful overall result could be established.

Todd's value for the distance of the Sun was 148 098 000 kilometres, when calculated with the modern value for the Earth's radius. He cautioned though that his results could not be 'regarded as definitive' for he had to adopt provisional values for the longitude of some of the stations.

Analysing the hundreds of images taken by the US expeditions was always going to be a difficult task. A lack of funds made the job even tougher.

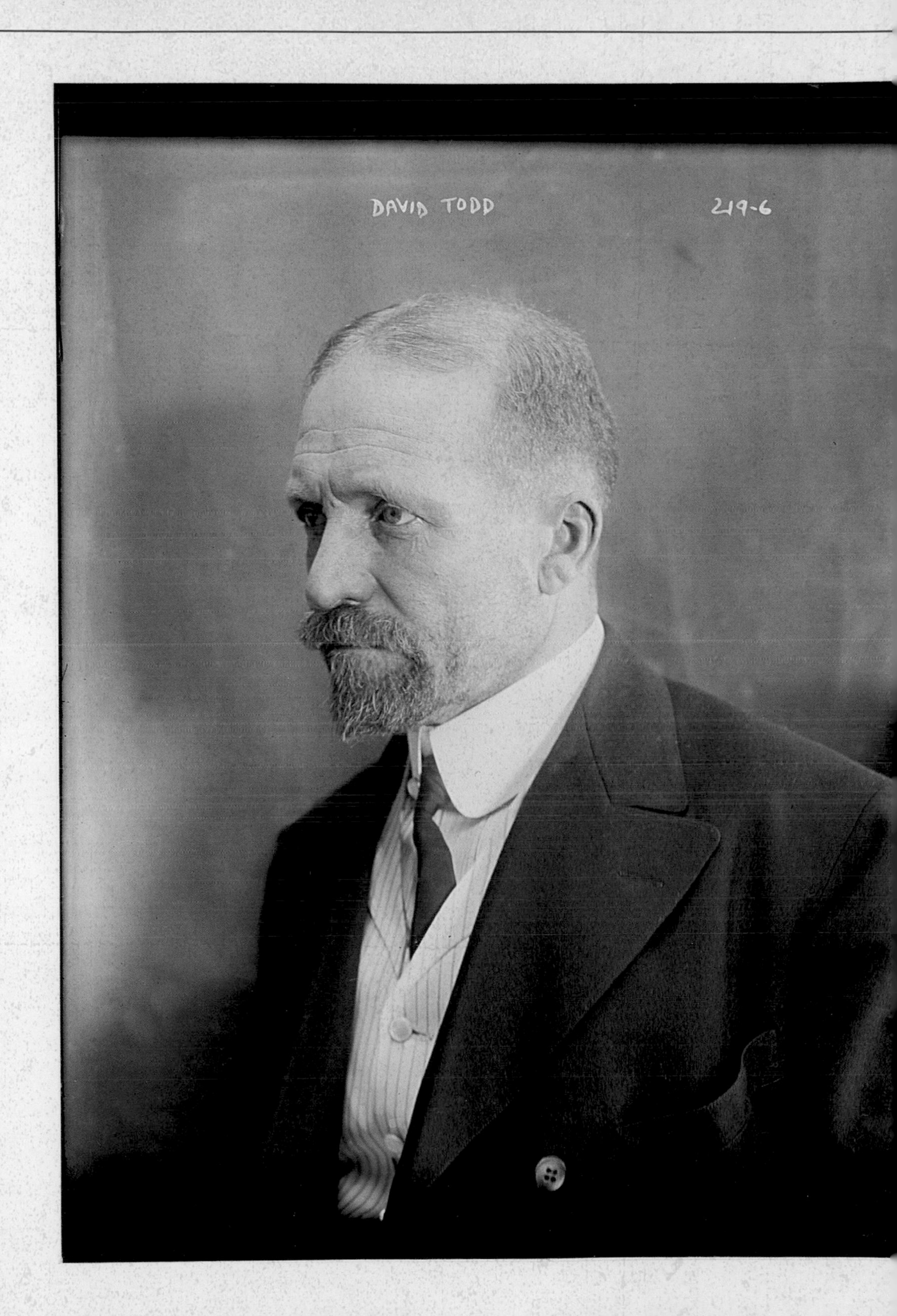

David Todd used photographs taken by the American teams in Tasmania to make a new calculation of the Sun's distance. He later moved to Amherst College, Massachusetts, where he became a professor of astronomy and director of the college's observatory. Library of Congress Prints and Photographs Division Washington, DC

Charles W Raymond led the second American expedition to observe the 1874 transit from Tasmania. Raymond was a captain in the US Army and a principal assistant professor of natural and experimental philosophy at the US Military Academy, West Point. Anchorage Museum, B1982.19.50

Two days before the transit Raymond issued detailed instructions to his party, clearly showing his military training, on 'the plan of operations' to follow on the day. They concluded with the exhortation:

> The chief astronomer reminds all who are connected with this party of the vast importance of the determinations to be made, and of the great responsibility which rests upon them. He expects from each strict attention to the duty confided to him, so as to insure [sic] the rapidity and accuracy which the preliminary drills have shown to be attainable.

As in Hobart, on the morning of 9 December, the day of the transit, the weather was grim and threatening. Whenever there was an opportunity, the photographers took photos. Altogether, 55 full Sun images were taken and 77 smaller ones during the period between third and fourth contacts. Although the Sun was partially covered by clouds in some of these photographs, they made a significant contribution to the total number of useful photographic plates obtained by all the American parties.

Raymond managed to observe the internal contact at egress in spite of 'light clouds' drifting over the Sun at the time. He notes that the phenomena at contact were just like those he had observed with the artificial transit during training at the Naval Observatory in Washington, although 'The cloudy link between the planet and the sun's limb, just before contact, seemed much more confused'.

At the end of his report on the transit observations at Campbell Town, Raymond acknowledges the 'generous hospitality' accorded to him and his party 'by the people of Tasmania and the neighboring British colonies of Australia'. He singles out a few people by name, including Dr Valentine, owner of The Grange, saying, 'The facilities furnished by him contributed much to the success of our labors'.

Surprisingly, almost one and a half centuries after the event and the departure of Captain Raymond's party, remains of the American visit to Campbell Town can still be seen. The most obvious remnant is the equatorial house. It has become a summer house in a slightly different position to its original and the rotating dome has been replaced, but the steel ring that was at the base of the dome is still in place at the top of the walls. Some of the timber parts of the building have been inscribed with roman numerals, clearly to assist in assembling the building.

The equatorial house set up by the American party led by Charles Raymond at Campbell Town, northern Tasmania, as it appears today. Tests on the timber walls have shown they are made of sugar pine from the United States. Photo: courtesy of Martin George

Transit of Venus - Dec. 9, 1874 - 1874.

To Mr. Alfred B. Biggs

From his friend

Chas. W. Raymond

Capt. of Engineers, U.S. Army

Chief Astronomer.

CLOUDS CLEAR IN NEW ZEALAND

New Zealand was a particularly favoured location for transit of Venus observations in 1874, with American, British, German and local observing teams. The American party was led by the professor of astronomy at Hamilton College, New York, Christian Heinrich Friedrich Peters.

The *Swatara*, which had carried all the American southern hemisphere parties, reached Bluff Harbour at the southern end of the South Island of New Zealand on 16 October 1874. The party's second-in-command, Lieutenant Edgar Wales Bass, had arrived a few weeks earlier in order to find a suitable observing location. Based on advice from local scientists, he selected the elevated town of Queenstown on the shores of Lake Wakatipu.

The party and its many boxes of equipment then began the difficult journey from Bluff Harbour to Queenstown. Travelling by train, wagon (equipment) or stage-coach (people) and finally, steamship, they reached Queenstown six and a half weeks before the transit. There they chose a site on Melbourne Street – now occupied by the modern Millennium Hotel – near the town centre as a suitable observing location.

One of the main tasks during those weeks before the transit was to establish the location of the Queenstown observing station by astronomical observation. In his report Professor Peters singles out Edward White at Melbourne Observatory for publishing a list of suitable southern stars from which he could choose 44 for his party's observations.

To establish the longitude, Peters and his party made telegraphic exchanges for time with a variety of places including the *Swatara*, when it was located in Port Chalmers near Dunedin. The ship's navigator carried one of the 20 chronometers on board to the local telegraph office and sent signals by tapping every five seconds. A similar set of exchanges were made with the ship that carried the German party to Auckland on the North Island and with the British party at Burnham, near Christchurch on the South Island.

The final determination of longitude depended on linking Wellington in New Zealand with Sydney. A cable was laid about a year after the American party

A print of one of the full disc photographs taken by Raymond's party during the 1874 transit. It is inscribed 'To Mr. Alfred B. Biggs, From his friend Chas. W. Raymond, Capt. Of Engineers US Army, Chief Astronomer'. Biggs was an amateur astronomer who had assisted the Americans at Campbell Town. Queen Victoria Museum and Art Gallery

left Queenstown and in March 1876 this was used by the New Zealanders and Henry Chamberlain Russell of Sydney Observatory to establish a longitude difference between the two cities. Peters used this value for his final determination of the longitude of the Queenstown site.

Of the day of the transit, Peters writes, 'Daylight came on gloomily, and with the sky overcast; it augured badly for the now approaching event. Telegrams from all parts of the island indicated that the cloudy cover was spread over a wide area'. Fortunately, the clouds cleared, at least for the beginning of the transit, and Peters successfully observed the two contacts during ingress. The party took a total of 237 photographs, the last of these taken 16 minutes before the expected start of egress. In his summary of the observations at Queenstown Peters notes that:

> The American party at New Zealand may be said to have been particularly favoured by the heavens. None of the other observers on this island have seen the least of the transit – clouds seeming to have overhung the whole area. I received telegrams in the forenoon such as these: Rain at the Bluff, rain at Clyde, overcast at Dunedin, overcast at Christ church, etc. It seems that we escaped disappointment by being at a greater elevation above the sea.

BRITISH IN KERGUELEN ISLAND

Father Stephen Joseph Perry was in charge of the British expedition to Kerguelen Island in the southern Indian Ocean. Trained by the Jesuits as a priest and as a scientist, Perry was in charge of the observatory at Stonyhurst College in Lancashire, England. Two Royal Navy ships, HMS *Volage* and HMS *Supply*, carried the expedition and its 600 cases of equipment to the desolate island of Kerguelen. Perry was plagued by sea-sickness during the voyage, which was a rough one.

On the day of the transit, a small cloud in the wrong spot in the sky frustrated the attempt to view the first internal contact, but the observations of the third and fourth contact were successful. Perry took photographs with a Janssen apparatus at the times of the contacts and with ordinary photographic plates in between times.

As with the other transit of Venus parties, establishing the location of the observing site, especially its longitude, was crucial. At Kerguelen this task was made exceptionally daunting by poor weather. The method involved making 100 double observations of the position of the Moon together with timing 30 passes by the Moon across the meridian, the imaginary overhead line crossing north to south. Up to the date of the transit the party had only observed about five passes across the meridian. Perry asked for more time. The ships' captains agreed to an extension of the expedition's stay, though, as Perry recorded, at a serious cost: 'Calculations were made on both sides, the question of provisions was weighed against that of lunar observations, and the balance struck was that observations may be continued till the end of February, even though it was necessary to put all upon half-rations'.

On the basis of the success at Kerguelen, Perry was put in charge of an expedition for the 1882 transit, this time to Madagascar. Then he went on a series of eclipse expeditions, to the West Indies in 1886, to Russia in 1887 and, finally, in 1889 to the Salut Islands, off the coast of French Guiana in South America. While there, he visited the French penal colony of Devil's Island, caught a serious infection and died on board ship.

The United States also sent an expedition to Kerguelen Island. This photograph depicts one of the American team at Kerguelen Island making magnetic field measurements. Photo: The James Melville Gilliss Library, US Naval Observatory

BRITISH IN HAWAII

The British transit of Venus expedition to Hawaii arrived at Honolulu on board HMS *Scout* on 9 September 1874. Expedition members were welcomed by King David Kalakua, who had been elected earlier that year to the position of monarch of the then still independent Hawaii. Three separate observing sites were established: at Honolulu, at Kailua on the Big Island of Hawaii and at Waimea on the island of Kauai.

The main observing site was set up in Honolulu under the charge of the expedition leader Captain George Lyon Tupman of the Royal Marines. The site, near the intersection of Punchbowl and Queen streets, was fenced off and the instruments were located inside.

On 8 December the weather was conducive to observations of the transit with a clear sky for most of the day and only a light wind. The local people were greatly interested in the activities of the astronomers. Tupman writes, 'Every tree in the neighbourhood of the inclosure, and every roof commanding a partial view within, bore its living freight'. Others lined up at the entrance hoping in vain to be allowed inside to look at the event through the telescopes.

Tupman began his observation of the transit using a spectroscope on one of the smaller telescopes. With this device that spreads out sunlight into its component colours he could see Venus as it approached the Sun against the backdrop of the thin outer layer of the Sun's atmosphere, called the chromosphere. Once he had seen Venus first touch the Sun at first contact, Tupman switched to a 6-inch (15-cm) lens telescope and began measuring Venus as it moved onto the Sun. Although there was no sign of the black-drop or similar effect, Tupman thought that he may have missed the all-important internal or second contact possibly while focussing the telescope and he was 'much disheartened at the unsatisfactory nature of the observation of the contact'.

Although only the beginning or ingress of the transit was visible from Hawaii, after the first two contacts Tupman still had time to take some further measurements before sunset. As the Sun and Venus set in the west the expedition team took stock of the observations that they had made including taking 60 photographs and then, before talking to each other, they wrote their own reports of the event.

On the homewards voyage Tupman planned to compare the time on the expedition's accurate clocks or chronometers with the time at San

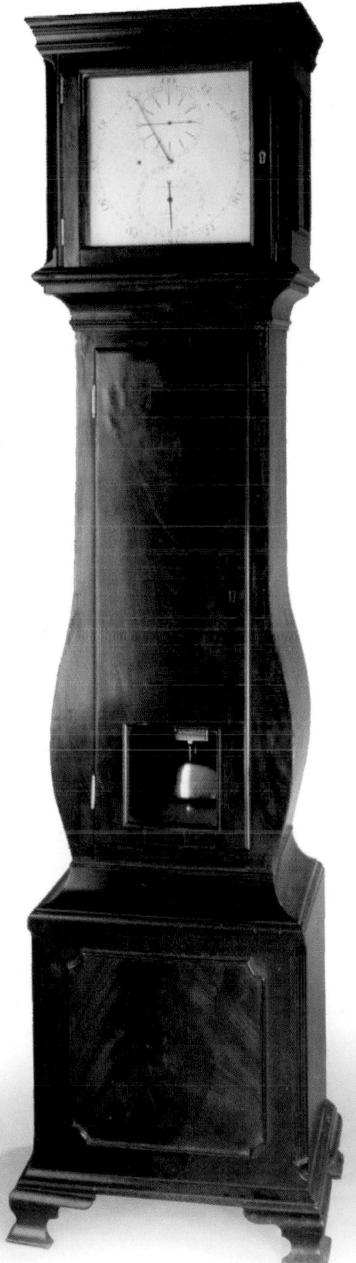

Astronomical regulator clock used by the British transit of Venus expedition at Kailua, Hawaii. The English chronometer-maker Thomas Earnshaw originally made this clock for use by Captain George Vancouver during his survey of the northwest coast of North America in the early 1790s. The clock was sent out to Vancouver in the care of a young astronomer William Gooch who was murdered in Hawaii during the voyage. Collection: Powerhouse Museum, Sydney.

Francisco in order to establish the longitude of the observing site. Unfortunately, contrary winds pushed the ship to colder latitudes, disturbing the timekeeping of the chronometers and extending the duration of the voyage. Despite this, on arriving at the US Navy Yard in San Francisco on 9 April 1875, the team still determined the time difference between the two cities. Then the longitude of the Honolulu observing site was established by the time difference and the longitude of San Francisco that was supplied by Professor George Davidson of the US Coast Survey, who had been in charge of the American observing team at Nagasaki.

At Kailua, George Forbes, the 25-year old Professor of Natural Philosophy at Andersons College in Glasgow, Scotland, was the leader of the British observing team. Forbes and his team were accommodated in the roomy palace of Princess Ruth Ke′elikōlani that had plenty of space for their instruments in its grounds. One of these instruments was an astronomical regulator clock originally built for the explorer George Vancouver and now on display at Sydney Observatory. On the afternoon of the transit clouds disrupted observations, though some views were obtained about halfway between first and second contact. The illumination of the outer edge of Venus while still off the Sun was clearly seen with the northern part of the planet brighter.

ANALYSING THE BRITISH AND COLONIAL RESULTS: 1874

Waiting for contact at Honolulu in 1874 where Captain GL Tupman was in charge of the British observing party. *The Illustrated London News,* 23 January 1875. Brian Greig collection

CAPTAIN GL TUPMAN MADE the most detailed treatment of the various 'British' observations taken around the world. In addition to those made by the official British Government expeditions, he used observations from the Cape of Good Hope, India, Australia and elsewhere. The Australian observations included those of the Sydney, Melbourne and Adelaide Observatories and that of John Tebbutt at his observatory at Windsor.

Tupman, a British military officer, had an interest in astronomy and was one of the first to try to estimate the velocity with which meteoroids hit the Earth's atmosphere. He had a major role in the British observations of the 1874 transit, starting as an instructor for the observing parties being trained at Greenwich prior to their departure to their various observing locations. Tupman was in charge of the observing party that travelled to Honolulu in the Sandwich Islands (Hawaii).

After Tupman returned from Hawaii, the Astronomer Royal, George Airy, asked him to stay at Greenwich to analyse the observations submitted by the various British observing parties. He spent four years on this analysis. According to his obituary published in 1923 in the *Monthly Notices of the Royal Astronomical Society*, Tupman did the work *con amore*, which literally means 'with love', but can be translated as 'with zeal'.

The analysis was not easy. Many of the timings were of necessity inconsistent and inaccurate, so that Tupman had to carefully interpret the description given by each observer. In addition, even before tackling the timings, he had to establish the longitude of each observing site from the large number of relevant measurements that had been made. The writer of Tupman's obituary in the *Monthly Notices* says that he had heard Tupman 'lament that the conditions of contact in the actual transit were so much less satisfactory than in the case of the clockwork model used for training the observers. In the latter case the accordance between observers was all that could be desired'.

Tupman treated the observations of ingress and egress separately instead of following Halley's suggested method of estimating the shift in the position of Venus' track across the Sun from the duration of the transit at each observing location. He noted that using Halley's method of durations was only appropriate when the contact times could be recorded more accurately than the coordinates of the observation sites. By 1874, this was no longer the case. Improvements in surveying instruments since the previous transit over a century earlier meant that most sites had accurately measured latitudes and longitudes, while sometimes even observers next to each other could only agree in their timings of the contacts to the nearest 30 seconds. Treating ingress and egress separately gave Tupman double the number of observations to use in his calculations and yielded two independent results to compare with each other instead of just one.

Though consistent results from ingress and egress were obtained, Tupman found that the 1874 observations could be satisfied by any value for the Sun's distance between 148100000 kilometres and 149200000 kilometres. This wide range of over a million kilometres was disappointing to Tupman and his contemporaries because it was no improvement over the results of the transits a century earlier. From a modern perspective, it is concerning that this range does not include the modern value of 149598000 kilometres. Tupman does note, though, that some of the Sydney observations had an effect of increasing the calculated distance, that is, they pushed the calculations in what we now know is the right direction.

According to his obituary, Tupman did the work con amore, *which literally means 'with love'.*

Members of Tupman's party at Honolulu practised by viewing an artificial transit from this shed. *The Graphic*, 5 December 1874. Brian Greig collection

OBSERVERS IN THE COLONY OF NEW SOUTH WALES

In Sydney, the bustling harbour of the British colony of New South Wales, Australia, Henry Chamberlain Russell, the director of Sydney Observatory and the first Australian-born Government Astronomer in the state of New South Wales, had started planning for the transit in 1870. Russell felt that 'It was obviously for the honor of the Colony as well as for the advancement of Science, that the observations and photographs of the transit should be as complete as possible'.

Russell and his staff would observe from Sydney Observatory, using as their main instrument the 11½-inch (29-centimetre) lens telescope that Russell had obtained specifically for the transit. To reduce the chance of clouds disrupting the observations,

Henry Chamberlain Russell in a January 1898 photograph by the commercial photographer J. Hubert Newman. Russell served a number of terms as president of the Royal Society of New South Wales and in 1888 became the first president of the Australasian Association for the Advancement of Science. Three years later he was appointed as the first Australian-born Vice-Chancellor of the University of Sydney. Ill health forced his retirement as Government Astronomer in 1905. Collection: Powerhouse Museum, Sydney

The large-lens telescope in the south dome of Sydney Observatory, photographed in the 1870s. The photograph is from *Observations of the Transit of Venus, 9 December, 1874.* Powerhouse Museum Research Library

HC Russell's observations of the start of the transit. Photo lithograph from *Observations of the Transit of Venus, 9 December, 1874.* Powerhouse Museum Research Library

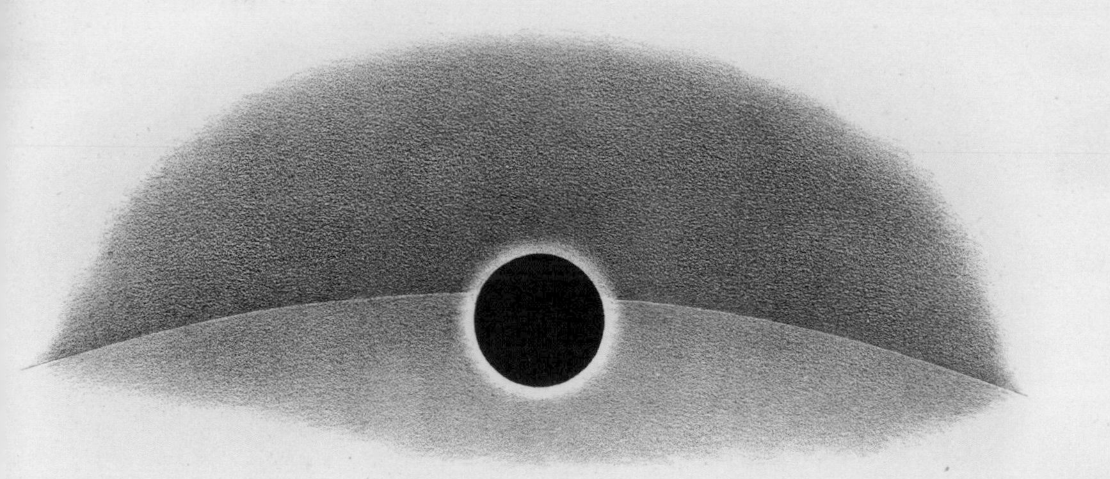

William John MacDonnell, one of the team observing from Eden, New South Wales, made this sketch of a 'shadowy envelope' surrounding Venus when it was two-thirds of the way onto the Sun's disc. Photo lithograph from *Observations of the Transit of Venus, 9 December, 1874.* Powerhouse Museum Research Library

Russell organised three other observing parties. He identified suitable sites at Woodford (in the Blue Mountains west of Sydney), Goulburn, further inland, and Eden, on the far south coast. Russell recruited some of the most prominent scientific people in the colony to staff these observing stations.

Russell placed great importance on the use of photography, as did most of the leaders of transit of Venus expeditions in other countries. They all hoped that this new technology would provide more accurate results for the Sun's distance than the old-fashioned technique of visual timings that had been used in 1761 and 1769. Photographs taken as Venus moved onto the Sun and as it moved off it were expected to eliminate the problems experienced by people such as Cook and Green at the previous transit due to the black drop effect. More importantly,

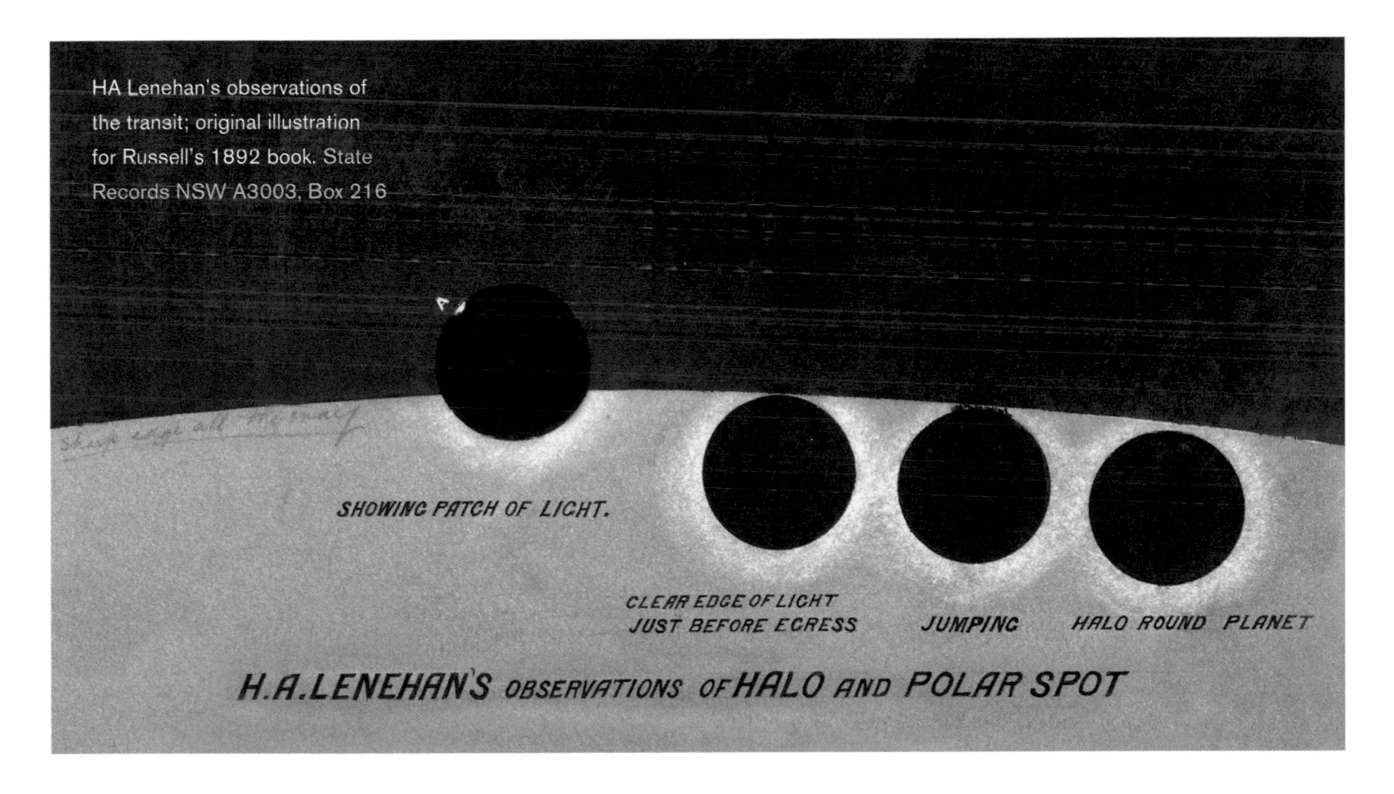

HA Lenehan's observations of the transit; original illustration for Russell's 1892 book. State Records NSW A3003, Box 216

photographs taken while Venus was moving across the Sun were expected to allow direct measurement of the different tracks across the Sun from widely separated observing stations.

Taking astronomical photographs today is a simple matter of attaching a camera, preferably digital, to the telescope. It was not so simple in 1874, when the wet collodion process was still in use. First a tent was placed inside the dome of the Observatory, with the telescope tube inserted into a light-tight opening so that a 'darkroom' was formed at the eyepiece or camera end of the telescope. A team of three was required, as Russell explained:

> One coated the plates and put it into the baths, of which four were used, fixed on a turn-table, so that by the time a plate had travelled around it was sensitized. The second worker took the plate out and put it into the camera, exposed it, and handed it to the third, who developed it and finished the picture …

> A view of the transit of Venus camp at Woodford in the Blue Mountains. The site, owned by the Sydney businessman Alfred Fairfax, later became the Woodford Academy and is now owned by the National Trust of Australia (New South Wales). Collection: Powerhouse Museum, Sydney

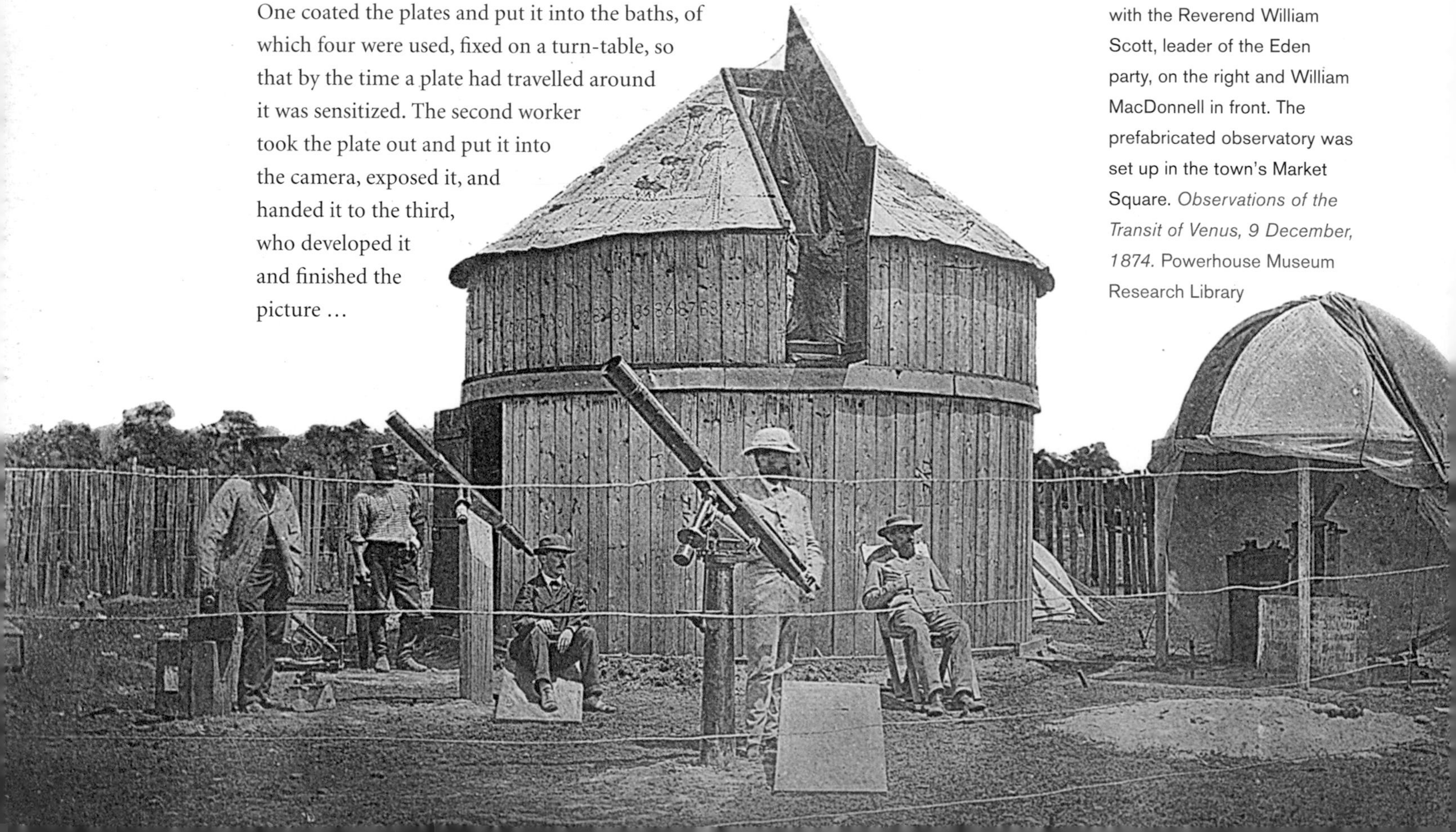

Waiting for the transit at Eden, with the Reverend William Scott, leader of the Eden party, on the right and William MacDonnell in front. The prefabricated observatory was set up in the town's Market Square. *Observations of the Transit of Venus, 9 December, 1874.* Powerhouse Museum Research Library

George Hirst's original illustration of the black drop phenomenon, an optical effect in which a black line is seen connecting Venus to the edge of the Sun. Drawn for HC Russell's 1892 book, *Observations of the Transit of Venus, 9 December, 1874.* State Records NSW A3003, Box 216

Waiting for the transit at Woodford in the Blue Mountains west of Sydney. When Russell was selecting sites for observing locations, meteorological observations suggested Woodford was likely to have the best viewing conditions. *Observations of the Transit of Venus, 9 December, 1874.* Powerhouse Museum Research Library

The 7-inch (18-centimetre) Merz lens telescope used at Eden, the lens of which still survives in the collection of the Powerhouse Museum.

This was the Eden party's main instrument. *Observations of the Transit of Venus, 9 December, 1874.* Powerhouse Museum Research Library

At the start of the transit, which began at 11.55 am, Russell observed with an eyepiece on the telescope, using an aperture to reduce the main lens to 5 inches (12.5 centimetres) in width and placing a piece of a special green glass before the eyepiece and a dark blue or neutral shade coloured glass against his eye. About ten minutes after first contact Russell could see the whole of Venus, and another ten minutes later he saw a halo around the part of the planet still off the Sun. His description is striking: 'It was very remarkable and beautiful, like a fringe of green light, through which the faintest tinge of red could be seen; it was brightest near the planet, and seemed to shade off to nothing, with a diameter estimated at one second of arc'.

By 3.55 pm, when Venus was about to leave the Sun's disc, the sky was still clear in Sydney, though clouds were forming in the west. Russell did not see the expected black drop at either ingress or egress, though there was some distortion just around the time of internal contact at egress. The halo was again seen on the part of the planet not on the Sun. Eventually it became irregular and brighter near the planet's north pole. Russell continued to see this white patch until only one minute before Venus completely left the disc of the Sun. One of Russell's staff, Henry Alfred Lenehan, also saw a similar halo through another, smaller telescope, except that the halo extended right

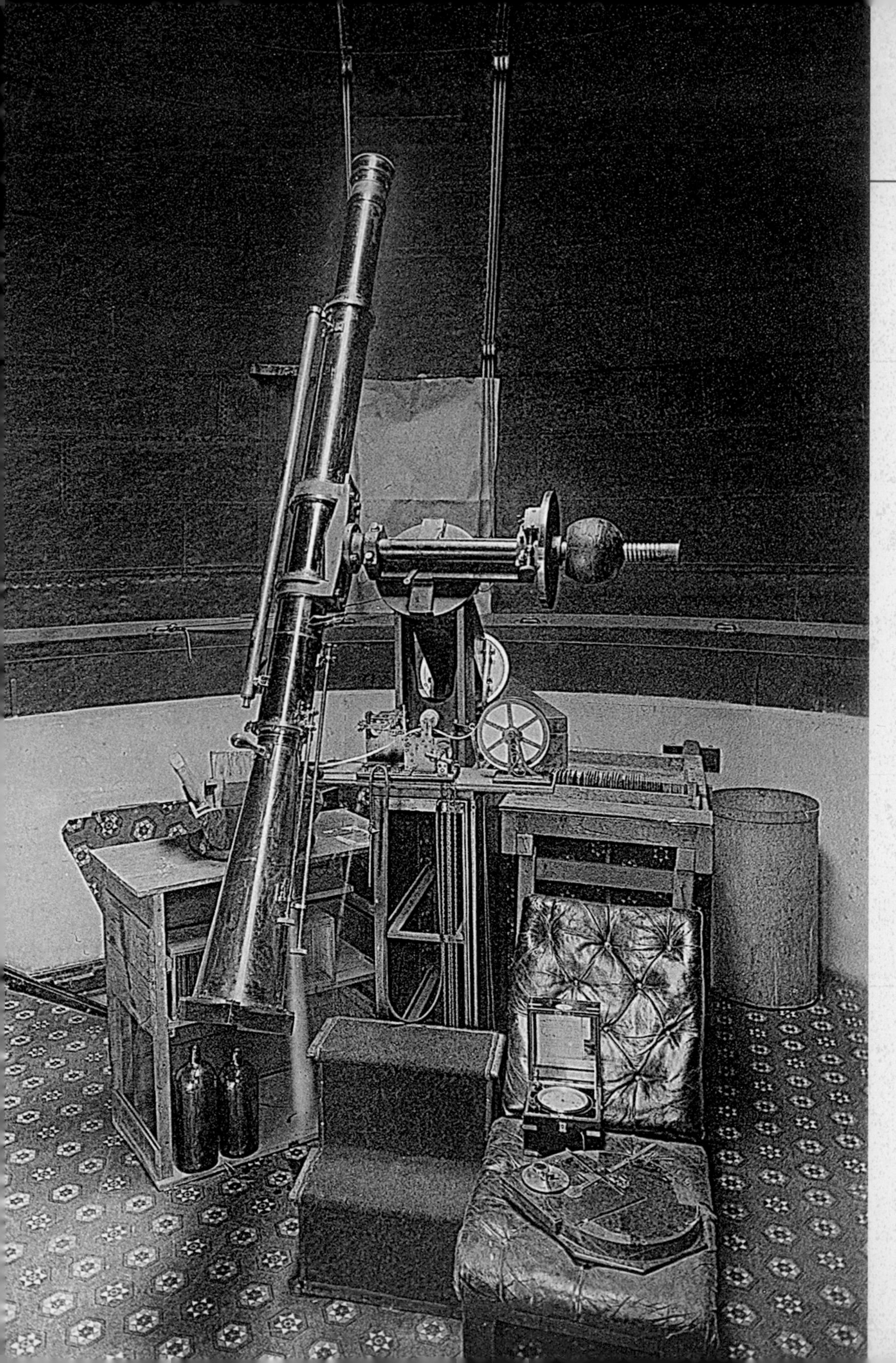

AT WOODFORD, THE PARTY'S main instrument was a photo-heliograph, a telescope specially designed to photograph the Sun, with a 4-inch (10-centimetre) objective lens and an enlarging lens to project a solar image of 10 centimetres diameter. It was built by the English optician John Henry Dallmeyer, who also made a number of identical telescopes for the five official British transit expeditions to Egypt, Hawaii, Rodrigues Island in the Indian Ocean, New Zealand and Kerguelen.

The photoheliograph was used together with a Janssen apparatus, known in France as a 'revolver photographique', to take images. With this ingenious device, developed by the French astronomer Jules Janssen, the astronomers could quickly take a succession of short-exposure photographs of the Sun on circular glass photographic plates. The images produced, each possibly the size of a passport photo (no one knows the exact size for sure), were arranged around the edge of the plate, and showed the same small section of the Sun's edge, with the movement of Venus on or off the disc captured as in a sequence of movie frames. This idea caught the attention of Britain's

< The photoheliograph used to photograph the 1874 transit at Woodford, shown here in the north dome of Sydney Observatory. The Janssen apparatus is on the observing couch on the right. *Observations of the Transit of Venus, 9 December, 1874.* Powerhouse Museum Research Library

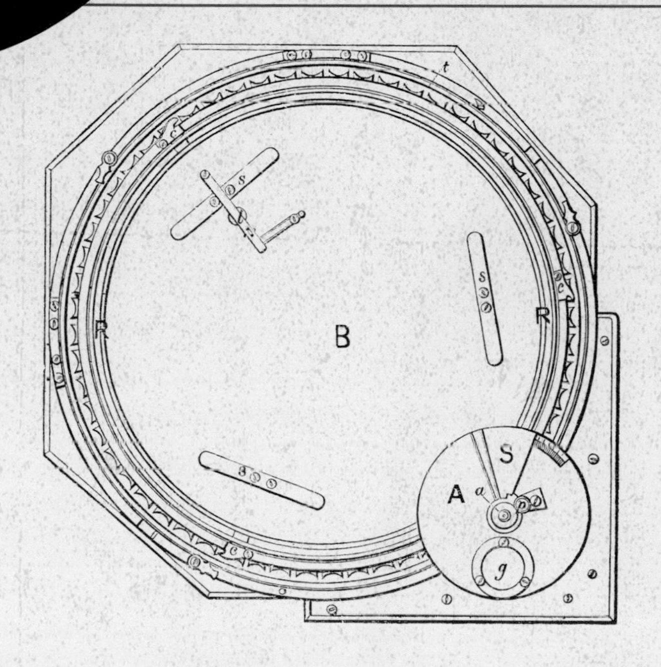

English astronomer Warren De La Rue's design for the Indian Janssen apparatus, which was probably the same as the one sent to Sydney Observatory. *Monthly Notices of the Royal Astronomical Society* May 1874. Sydney Observatory Collection

Astronomer Royal, Sir George Airy, but he wanted a different version. On 25 October 1873 he wrote to the English astronomer and expert on astronomical photography Warren De La Rue, 'Janssen has not the least idea of mechanics, and having got his principle, I throw him overboard'.

Dallmeyer built five Janssen apparatus for various English parties using his photoheliograph for recording the 1874 transit. De La Rue designed a slightly different version that could take 60 images of the Sun on one sensitised glass plate for use at an observatory in India and a smaller one that could take only 20 images for Melbourne. The one used at Woodford seems identical to the larger De La Rue version built for India and so is certainly designed by him. Now on display at Sydney Observatory, it is believed to be the only one of the British versions of the Janssen apparatus to have survived.

After the transit Jules Janssen suggested that his photographic revolver could have a variety of uses, such as studying the movement and flight of animals. He was acquainted with the Lumière brothers, Auguste and Louis, who developed the first movie camera. It seems likely that the Janssen apparatus was a forerunner of that camera. In fact, Louis Lumière filmed Janssen's arrival at a French photographic society meeting on 15 June 1895, only two months after the recording of the first movie footage.

The owner of the Woodford observing site, Alfred Fairfax, made this spectacular illustration of what he had seen through his own telescope: a halo around part of Venus as the planet moved onto the Sun. Fairfax noted that the illustration was not to scale and the halo was in fact much narrower. State Records NSW A3003, Box 216

around the planet, was about one third as wide as the planet, and was greenish-yellow in colour, with the outer edge shaded orange.

At Eden, the team of astronomers were hampered by cloud, but still managed some good observations and took more than 50 photographs. A few minutes before internal contact, the leader of the team, the Reverend William Scott, saw the planet outlined by a ring of light, which he attributed to light being scattered by the atmosphere of Venus.

At Woodford, the amateur astronomer George Hirst was the only one of the team to see the black drop. His task was to take photographs using the Janssen apparatus attached to the photoheliograph. To do this he had to point the telescope using a finder-telescope consisting of a low-quality lens of 1½-inch (4-centimetre) width as the objective, and an eyepiece. When he had placed the fifth Janssen plate into the apparatus, he glanced into the finder before beginning the exposure and saw Venus 'connected with the limb by a narrow line intensely black with an ill-defined edge' – a classic description of the black drop.

Meanwhile, at the Goulburn observing station, Archibald Liversidge, a professor at the University of Sydney, using a telescope with a

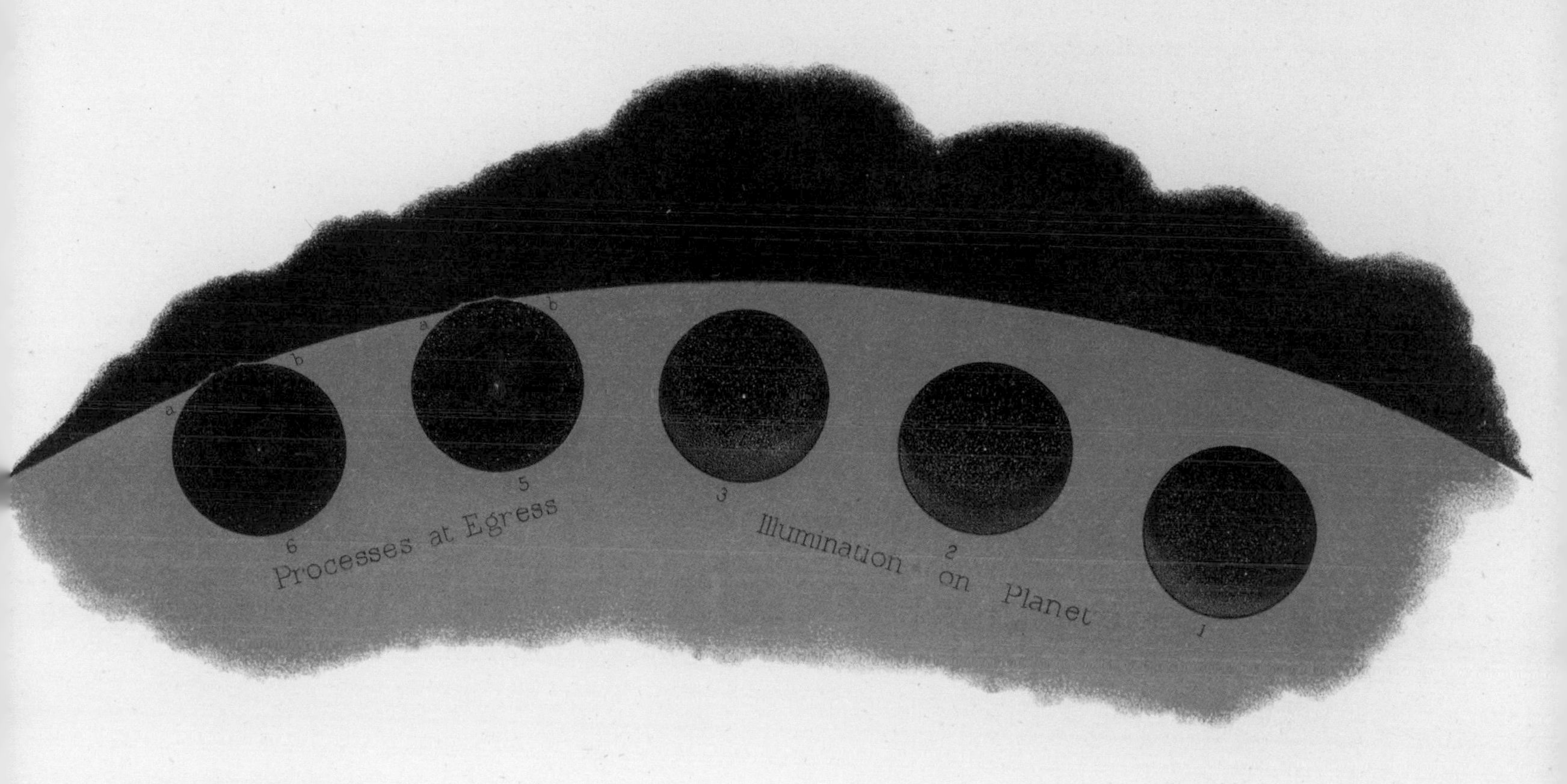

Transit of Venus 1874

Professor Liversidge's Observations

Archibald Liversidge made this sketch of his observations at Goulburn, New South Wales. Liversidge later established the forerunner of the Powerhouse Museum, Sydney. Photo lithograph from *Observations of the Transit of Venus, 9 December, 1874.* Powerhouse Museum Research Library

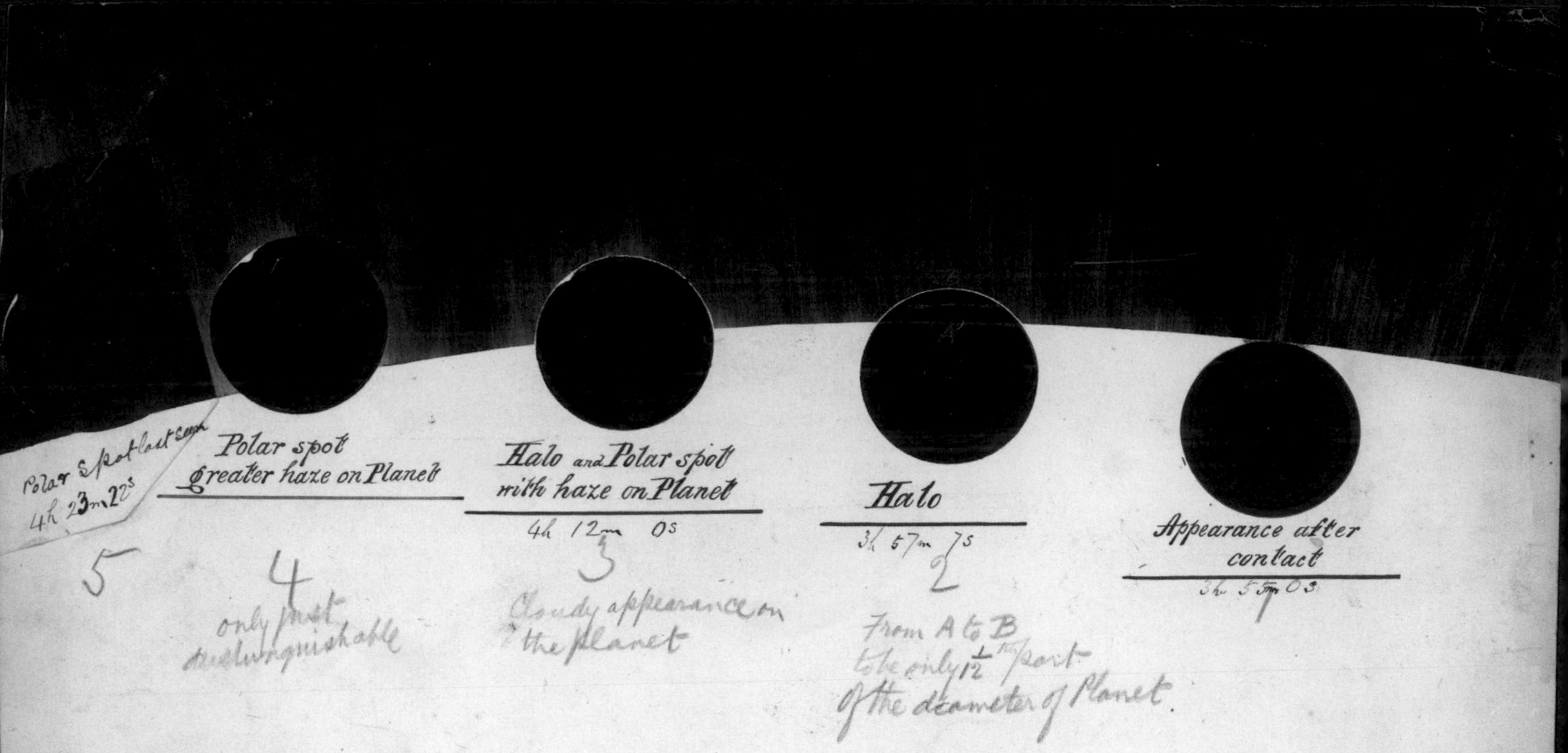

Polar spot last seen
4h 23m 22s
5

Polar spot
Greater haze on Planet
4
only just
distinguishable

Halo and Polar spot
with haze on Planet
4h 12m 0s
3
Cloudy appearance on
the Planet

Halo
3h 57m 7s
2
From A to B
to be only $\frac{1}{12}$th part
of the diameter of Planet.

Appearance after
contact
3h 55m 0s
1

— Mr Russell's observations ~~at~~ Egress

No 2

< HC Russell's observations at the end of the 1874 transit as seen from Sydney Observatory; original illustration for his 1892 book. State Records NSW A3003, Box 216

3¼-inch (8-centimetre) diameter lens, was surprised by the absence of the black drop, feeling that he had not made sufficiently detailed observations as Venus crept onto the Sun. Later, when the planet was leaving the Sun, it appeared to him that Venus was pushing the adjacent part of the solar limb before it.

In 1892 Russell published a beautifully illustrated book on the observations of the 1874 transit of Venus 'made at stations in New South Wales'. It is a compilation of reports and illustrations from the various observers.

A sketch of Sir George Airy, the British Astronomer Royal, published in the magazine *Vanity Fair,* 13 November 1875. Courtesy: Ian Ridpath

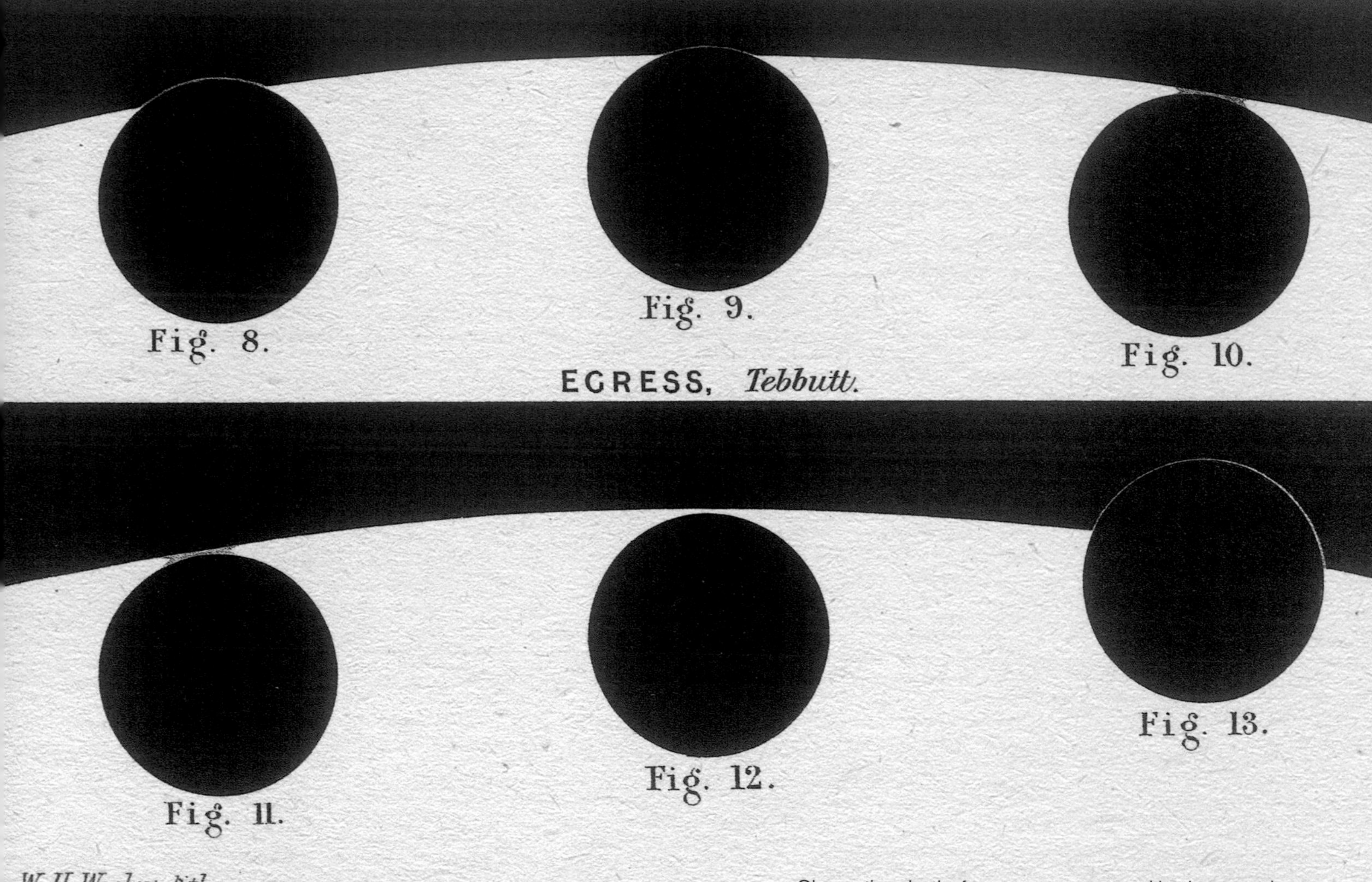

Observations by the famous amateur astronomer John Tebbutt of the transit of Venus in 1874, from near Sydney, Australia. Note that in spite of the placement of the word 'egress', Figs 8 to 12 all relate to ingress and only Fig. 13 to egress. That last image is a fine depiction of the ring of light scattered by the atmosphere of Venus. For his observations Tebbutt used a 4½-inch (11-centimetre) lens telescope with a diagonal eyepiece so that he could observe looking down instead of along the tube. *Memoirs of the Royal Astronomical Society 1882.* Brian Greig collection

THROUGH CLOUDY SKIES IN VICTORIA

Like a general before a battle, the Government Astronomer in the British colony of Victoria, Australia, Robert Lewis John Ellery, marshalled his team at the Melbourne Observatory, allocating duties and arranging the available instruments. The chief assistant, Edward J White, was to observe the external contacts and make measurements with the 'Double Image Micrometer' at the 8-inch (20-centimetre) lens Troughton & Simms telescope. By bringing two images into coincidence, the micrometer allowed small differences in angle to be measured, such as the angular distance between the centres of Venus and the Sun. After using the device, White was to race to a smaller telescope to observe the internal contacts while Ellery himself was to observe the same phases with the 8-inch. William Charles Kernot was to operate the Dallmeyer photoheliograph, and Joseph Turner was in charge of the famous Great Melbourne Telescope.

On the morning of the transit the sky was overcast with, according to Ellery, 'thunder, lightning and occasional rain'. At 11.36 am local time, though, Edward White managed to look through a break in the clouds and saw Venus already entering the solar disc.

While Venus was fully on the disc of the Sun, Ellery scrutinised the planet and its immediate surroundings for signs of an atmosphere or a satellite, but no such signs were observed. He did see that the outer part of the planet was coloured a very deep blue or indigo. A modern expert on optics, Dr Barry Clark, suggests that most likely the coloured fringe was a secondary spectrum effect – that is, the telescope lens was unable to focus all the colours to form a sharp image.

In case of poor weather, Ellery set up three observing stations in Victoria apart from the one at Melbourne Observatory. One of these was at Mornington, a small town about 50 kilometres south of Melbourne on the shore of Port Phillip Bay. In charge was Professor William Parkinson Wilson.

Despite poor weather, Wilson made some observations, noting, for example, as Venus was approaching internal contact, 'a small dark object flickering backwards and forwards between Venus and the edge of the Sun'. He was unable to time the internal contact at egress as he did not see a suitable phase for that occurrence.

Professor Wilson's observations of the transit had a tragic ending. He had been in ill health for some time,

THE GREAT MELBOURNE TELESCOPE

Melbourne Observatory used its biggest 'gun' to observe the transit: the largest fully steerable telescope in the world.

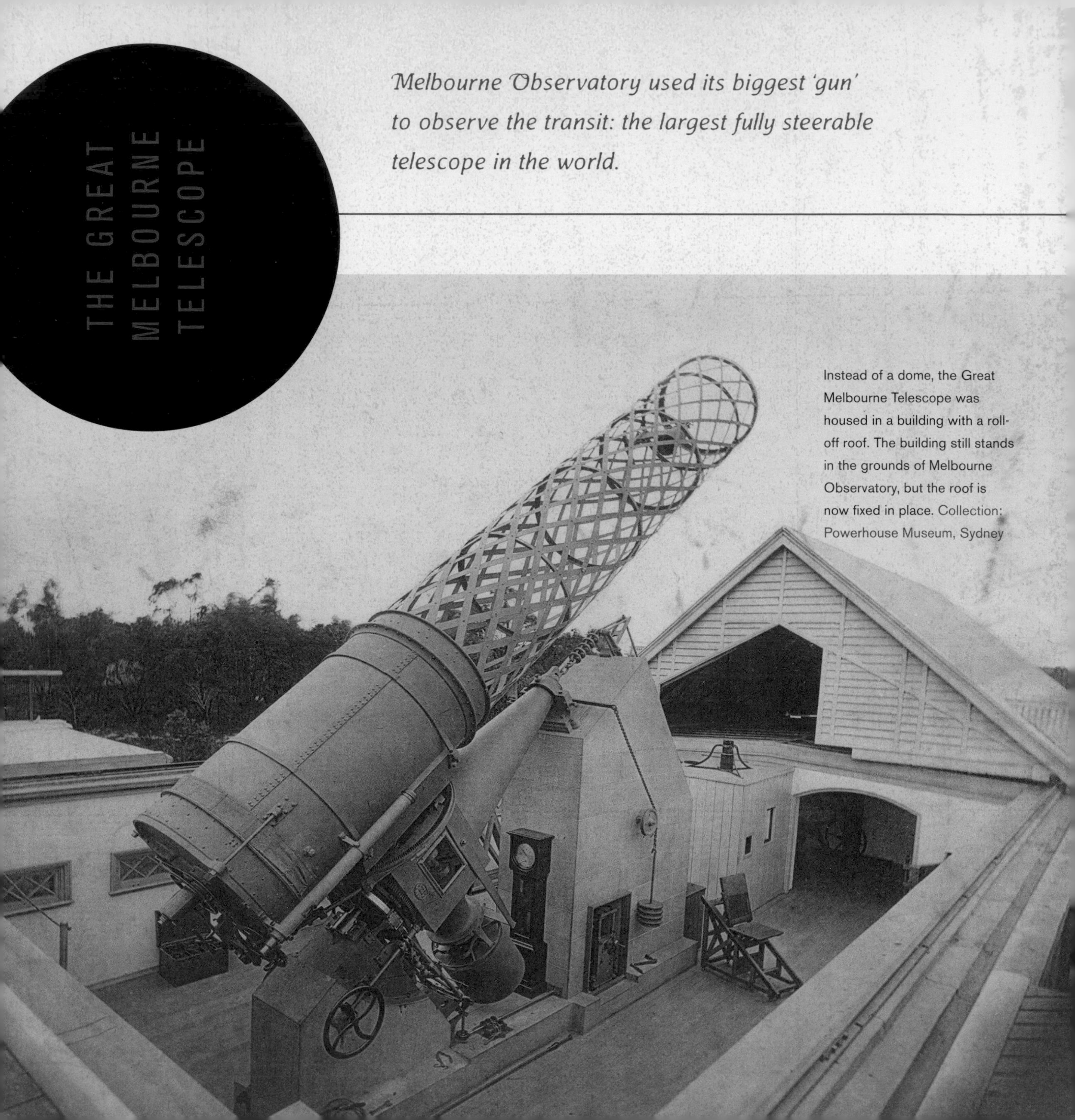

Instead of a dome, the Great Melbourne Telescope was housed in a building with a roll-off roof. The building still stands in the grounds of Melbourne Observatory, but the roof is now fixed in place. Collection: Powerhouse Museum, Sydney

MELBOURNE OBSERVATORY housed the largest fully steerable telescope in the world, the Great Melbourne Telescope. This huge instrument had arrived in Melbourne in December 1868 and was operational after many difficulties by the middle of the following year. The telescope's main project was to sketch the same nebulae or fuzzy objects that Sir John Herschel had sketched in South Africa earlier in the century. The comparison was to show if there had been changes in these objects in the intervening decades. This plan was not very successful as different telescopes and different observers made comparisons highly uncertain.

A century or so later the telescope had an interesting afterlife at Mt Stromlo Observatory near Canberra. There it became the basis of a modern automated telescope that was used to try to probe dark matter, one of the great mysteries of astronomy. Sadly, the telescope, together with others, was destroyed in the Mt Stromlo bushfire of 18 January 2003. The burnt remains have since been returned to Melbourne, where under the supervision of the Museum of Victoria, a group of keen volunteers from the Astronomical Society of Victoria have been cleaning and cataloguing the remaining parts of the telescope. Their aim is to rebuild the giant telescope at its original location and make it available for public observation.

The remains of the Great Melbourne Telescope in November 2004 after the Mt Stromlo fire. Photo: Nick Lomb

and after the transit complained about the heat and about being fatigued. Two days later his doctor was called by telegram to his Mornington home. Sadly, he died of a cerebral haemorrhage, a type of stroke, two hours before the doctor could reach him. Although what caused the stroke can never be known, it is reasonable to assume that the stress, excitement and exertion associated with the transit observations had contributed to the sad event. Like Chappe d'Auteroche in Mexico in the previous century, we can regard William Parkinson Wilson as a casualty of the transit of Venus.

Ellery set up another station at Glenrowan, a small town 230 kilometres from Melbourne. Here, the skies were once again grey. Between clouds, one of the team, James E Gilbert, noticed the illumination of the edge of Venus off the Sun or at least its northwest part. Later he saw a ligament between Venus and the edge of the Sun and managed to get a good timing of the instant that this 'ligament snapped suddenly'. As Venus crossed the solar disc, Gilbert saw the planet having a violet colour though with a central dark spot. He adds that during the whole time the surface of Venus 'appeared granulated, or, more correctly speaking, dappled'. The planet's egress could not be observed as the sky had become overcast, with rain and thunder.

> Melbourne Observatory photographed from Government House some time after 1874. The main building and associated telescope domes are on the left, while the Great Melbourne Telescope is on the right. In the last decades of the nineteenth century, Melbourne and Sydney Observatories were the two main astronomical and meteorological centres in Australia. Collection: Powerhouse Museum, Sydney

Robert LJ Ellery, the Victorian Government Astronomer, pictured in 1875. *The Australasian Sketcher* 17 April 1875. State Library of Victoria

—Melbourne. Observatory.—

THE PHOTOHELIOGRAPH

THE PHOTOHELIOGRAPH was another telescope used for the Melbourne Observatory observations. Built by the London maker John Henry Dallmeyer, this telescope had been purchased for £364 specifically to view the transit and had arrived at the observatory on 28 August 1874. Two new domes were built at the Observatory, one for the photoheliograph and the other for the 8-inch (20-centimetre) Troughton & Simms lens telescope. As at Sydney, a Janssen apparatus (see page 145) to allow quick consecutive images was also available for use with the photoheliograph. Designed by the English expert on solar photography, Warren De La Rue, this Janssen apparatus was constructed to be the same size as the normal 15-centimetre square photographic plate-holder for the telescope. This small size constrained the apparatus so that it could only take 20 images on one circular photographic plate instead of the 60 taken by the device used at Sydney.

Kernot, in charge of the photoheliograph, had a team of assistants to help sensitise the plates, load them into the apparatus and then develop them. One assistant kept the telescope pointing in the right direction as Kernot turned the handle of the Janssen apparatus at the rate of once a second. He knew the seconds because another assistant stationed in front of a box chronometer called them out. In this way Kernot and his assistants succeeded in taking 180 images of the Sun on at least nine plates while Venus was entering or leaving the Sun. Additionally, using the ordinary plate-holder they took 37 images of the whole disc in between the contact times.

The observers succeeded in taking at least 180 images of the Sun on at least nine plates while Venus was entering or leaving.

> The only surviving image taken by Melbourne's photoheliograph during the 1874 transit. The image has been converted to a positive one. The Science and Technology Facilities Council, UK

11
Melbourne

The final Victorian observing station was at View Hill, Sandhurst (now Bendigo), where Carl Moerlin was in charge. Clouds prevented seeing any of the ingress, Venus moving onto the Sun. Moerlin did, however, observe the planet as it neared internal contact at egress, when he saw 'a sort of triangular-shaped construction' between the edges of the planet and the Sun. The base of the triangle was on the Sun while the apex was on the planet, except when 'every once in a while' the apex jumped off it. He managed to obtain a good timing for the contact, but afterwards the clouds became thicker and he could not make any further measurements.

Sketches showing Victorian observations of the 1874 transit of Venus. *Memoirs of the Royal Astronomical Society* 1882. Brian Greig collection

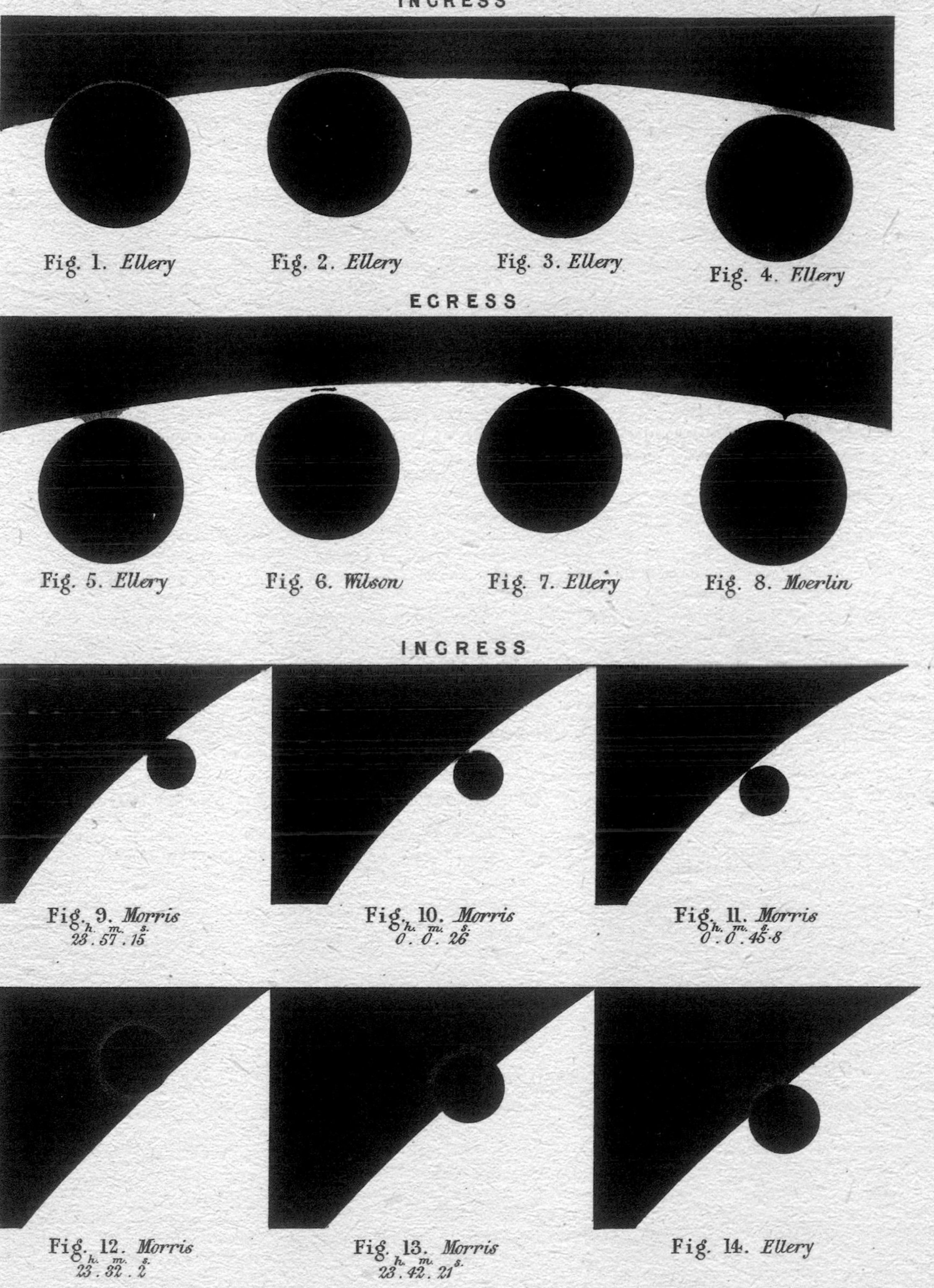

< Robert LJ Ellery ran the Williamstown Observatory in Victoria (pictured in 1862) from soon after its establishment in 1853. Ellery remained in charge after the observatory moved to its present site in Melbourne in 1862. *The Illustrated Melbourne Post* 15 November 1862. State Library of Victoria

THE 1882 TRANSIT

The next transit was to occur on 6 December 1882. By that time, scientific interest had somewhat waned because scientists had developed other methods of gauging the Sun's distance. The most important of these was the Mars parallax method. This involved measuring the shift in the position of the planet in the sky when seen from two places separated by the width of the Earth or from one place in the morning and in the evening. In 1877 the Scottish astronomer David Gill made a good measurement of the Sun's distance from Ascension Island in the South Atlantic Ocean.

Nevertheless, in 1882 the British established observing teams at places around the globe such as Canada, Jamaica, Madagascar, and Australia. Altogether the British teams made 34 observations of Venus moving on to the Sun and 39 moving off it.

To measure longitude accurately, observers exchanged time signals by telegraph, as their counterparts had done in 1874. A major focus was on obtaining longitudes for places in Australia. Charles Todd headed this effort, with Russell in Sydney and Ellery in Melbourne assisting. An essential component of determining the Australian longitudes was the exchange of signals between Port Darwin (now Darwin) and Singapore, which already had a well-determined position. The observer at Singapore happened to be Captain Leonard Darwin of the Royal Engineers, the son of evolutionary scientist Charles Darwin. According to the Royal Astronomical Society in 1884, the results of eight nights of exchanging signals between the two Darwins were 'very accordant'.

America sent out eight teams for the 1882 transit, including one to South Africa and one to San Antonio, Texas, and with generally favourable weather collected 1380 measurable photographic plates of the transit (four times as many as eight years earlier).

The 1882 transit was visible throughout the United States and this time mostly from beginning to end. There was great anxiety though the night before the event for 'a large portion of the United States was shadowed with thick clouds, and rain and snow were falling upon the roofs of the observatories' (*The Sun*, New York NY, 7 December 1882, p1). Fortunately, by the start of the transit in the morning from New York and many other places there were sufficient gaps in the clouds for observations to begin.

During the transit thousands of members of the public stared at the Sun and Venus through smoked and coloured glasses. A few enterprising amateur astronomers set up telescopes in the

This engraving of children observing the transit of Venus appeared on the cover of the New York magazine, *Harper's Weekly,* on Saturday 28 April 1883. British and American painter John George Brown painted the original version. Brian Greig collection

parks and on the sidewalks of New York in order to collect dimes from those who were 'eager for a glance at the strange spectacle in the sky'. In other places throughout the country people had similar opportunities. In the small town of Meriden, Connecticut, a fire alarm bell was rung to signal the start of the transit and people could go to the grounds of the house of a local amateur astronomer, Rev JT Pettee, where seven telescopes were provided for public viewing. Some six thousand people availed themselves of the opportunity.

Perhaps the most important observations in the United States were made at the US Naval Observatory in Washington, DC. A reporter for *The Sun* described these observations in detail, concentrating on the observations of Professor Edgar Frisby with the great 26-inch (66-cm) telescope.

Around the time of second contact there was a loud noise inside the telescope dome as the eyepiece of a smaller telescope cracked due to the heat from the Sun. The reporter says, possibly slightly tongue-in-cheek, that 'The 26-inch object glass of the big telescope had been protected from the first, otherwise it would have concentrated heat enough in a few minutes to have cremated Prof. Frisby.'

For the 1882 transit, William Harkness was in

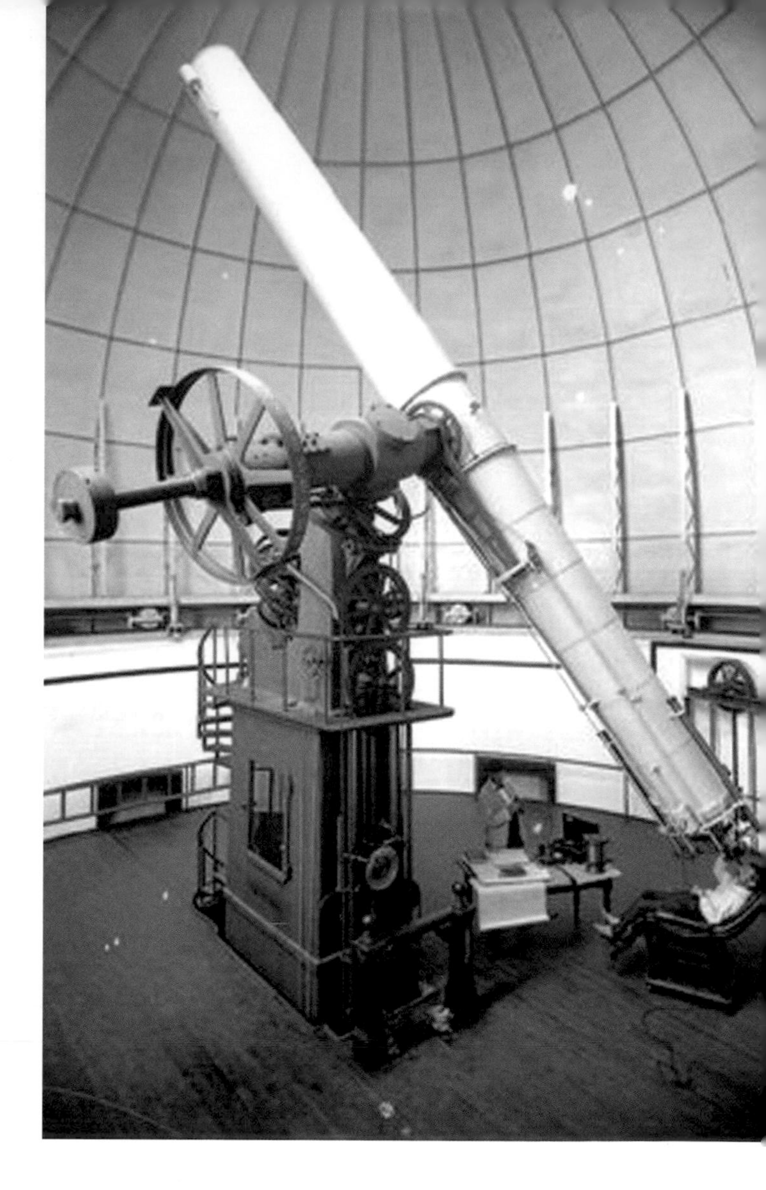

The famous 26-inch telescope at the US Naval Observatory with which the moons of Mars were discovered in 1877 and which was used for the observations of the 1882 transit of Venus. Photo: The James Melville Gilliss Library, US Naval Observatory

Professor Edgar Frisby, who was born in England and studied in Canada, observed the 1882 transit with the great 26-inch telescope at the US Naval Observatory. After spending a decade as an assistant astronomer at the observatory, he was appointed a professor of mathematics in the United States Navy. In the late 1880s, the first African American mathematics graduate student, Kelly Miller, studied advanced mathematics with him. Library of Congress, Prints & Photographs Division, photograph by Harris & Ewing LC-DIG-hec-15426

charge of both measuring the plates and analysing the results. In 1891, he published a result from the photographs taken at that transit of 148 785 000 kilometres for the distance of the Sun from the Earth, a value a little closer to the modern value than that of Todd from the photographs of the previous transit.

It soon became clear to astronomers that the parallax method applied to the minor planets – the thousands of small rocky objects circling the Sun in addition to the major planets – promised even better results than with Mars. The minor planet Eros, discovered in 1897, was ideal for the purpose. In 1900-01, Eros passed by the Earth at about a quarter of the Sun's distance. Observations then and at its even closer passage three decades later finally settled the question of the distance of the Sun and the scale of the solar system. With the availability of radar in recent times, all that was left to do was to slightly refine the value.

SPACE-AGE TRANSIT

2004

By the time of the next transit of Venus in 2004, the space age had arrived. In the intervening century astronomers and scientists had accumulated a vast body of knowledge relating to Venus and our solar system. Advances in technology meant that we were not only able to record the transit of Venus from space but that hundreds of amateur astronomers could produce their own stunning photographs.

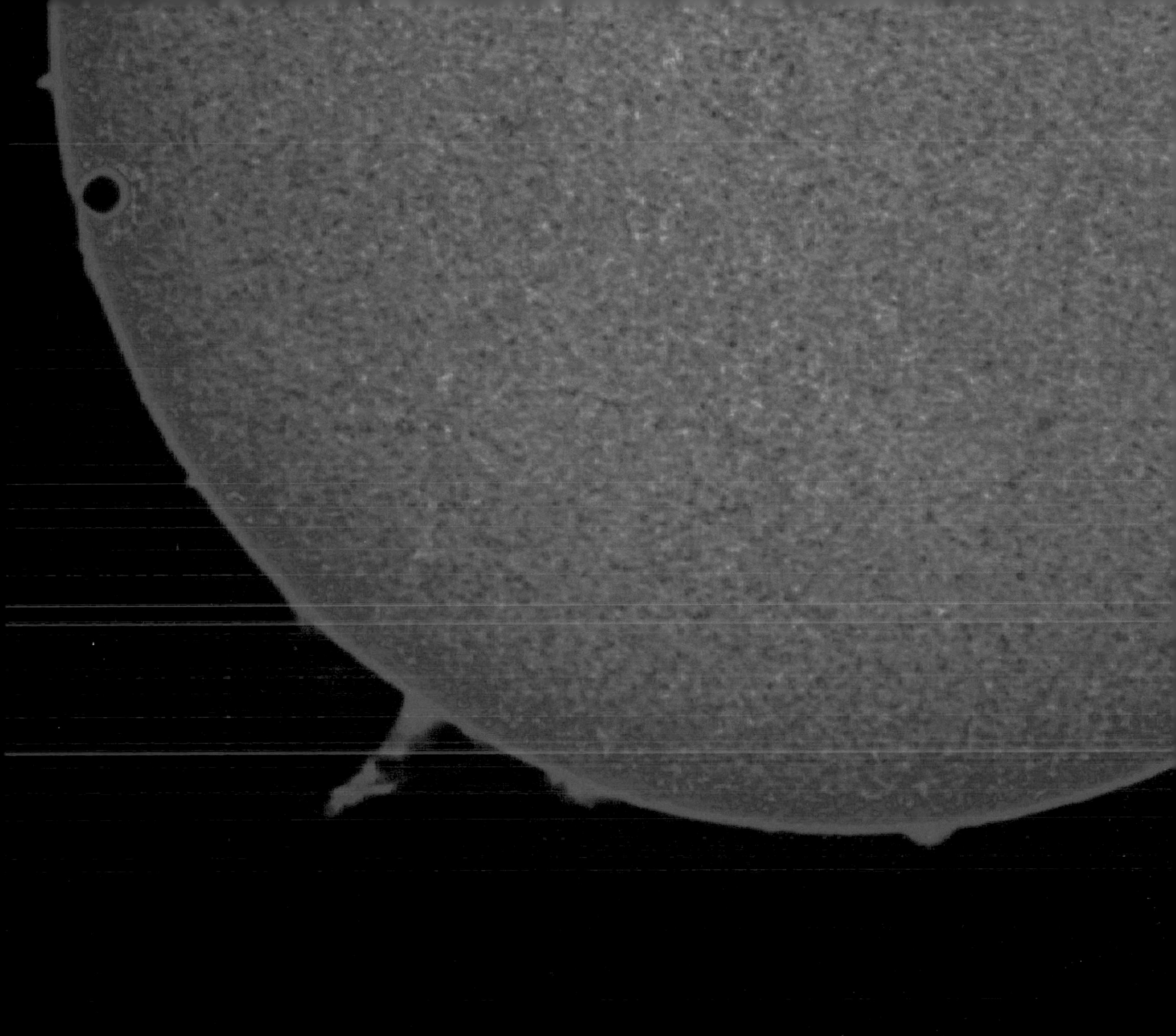

Geoff Wyatt's prize-winning image of the 2004 transit of Venus was taken in the north dome of Sydney Observatory through a special filter that only transmits the red light of hydrogen atoms. Photo: Geoff Wyatt, Sydney Observatory/Powerhouse Museum

Magellan produced hundreds of brilliant and detailed images of the planet's surface.

NASA's MAGELLAN RADAR imaging spacecraft was launched from the Space Shuttle *Atlantis* on 4 May 1989. The spacecraft spent four years in orbit around Venus between 1990 and 1994, giving us the first close look at the planet.

Magellan radar-mapped 98 percent of the planet's surface, collecting high-resolution gravity data and producing hundreds of brilliant and detailed images of the planet's surface. The $551 million mission generated more digital data than all previous US planetary missions combined. Magellan's innovative method of radar mapping, called synthetic-aperture radar (SAR), unveiled many of Venus' secrets, and we now know more about the geological and climatic conditions on Venus than ever before.

The mission came to a dramatic conclusion in October 1994 when Magellan was commanded to plunge into the planet's dense atmosphere. By crash landing on Venus Magellan collected data on the planet's atmosphere and on the performance of the spacecraft as it descended.

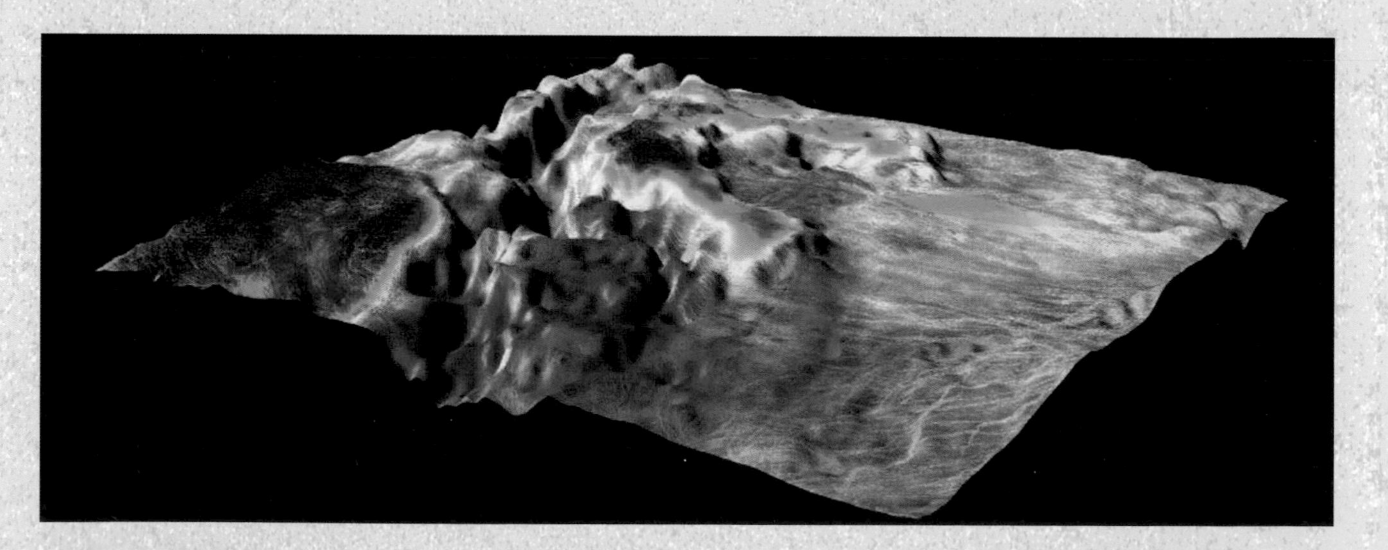

This perspective view of part of the surface of Venus, generated by computer from Magellan data, shows the boundary between the lowland plains and characteristic Venusian highland terrain in Ovda Regio, the western part of the great equatorial highland called Aphrodite Terra. NASA/PIA00311

The Magellan spacecraft's deployment from the shuttle *Atlantis*' cargo bay was captured by an astronaut with a hand-held camera pointed through the shuttle's aft flight deck windows. NASA P-34252BC

This hemispheric view of Venus was compiled mainly from Magellan images. The view is centred on 90 degrees east longitude, with the colours indicating elevation. NASA/JPL/USGS

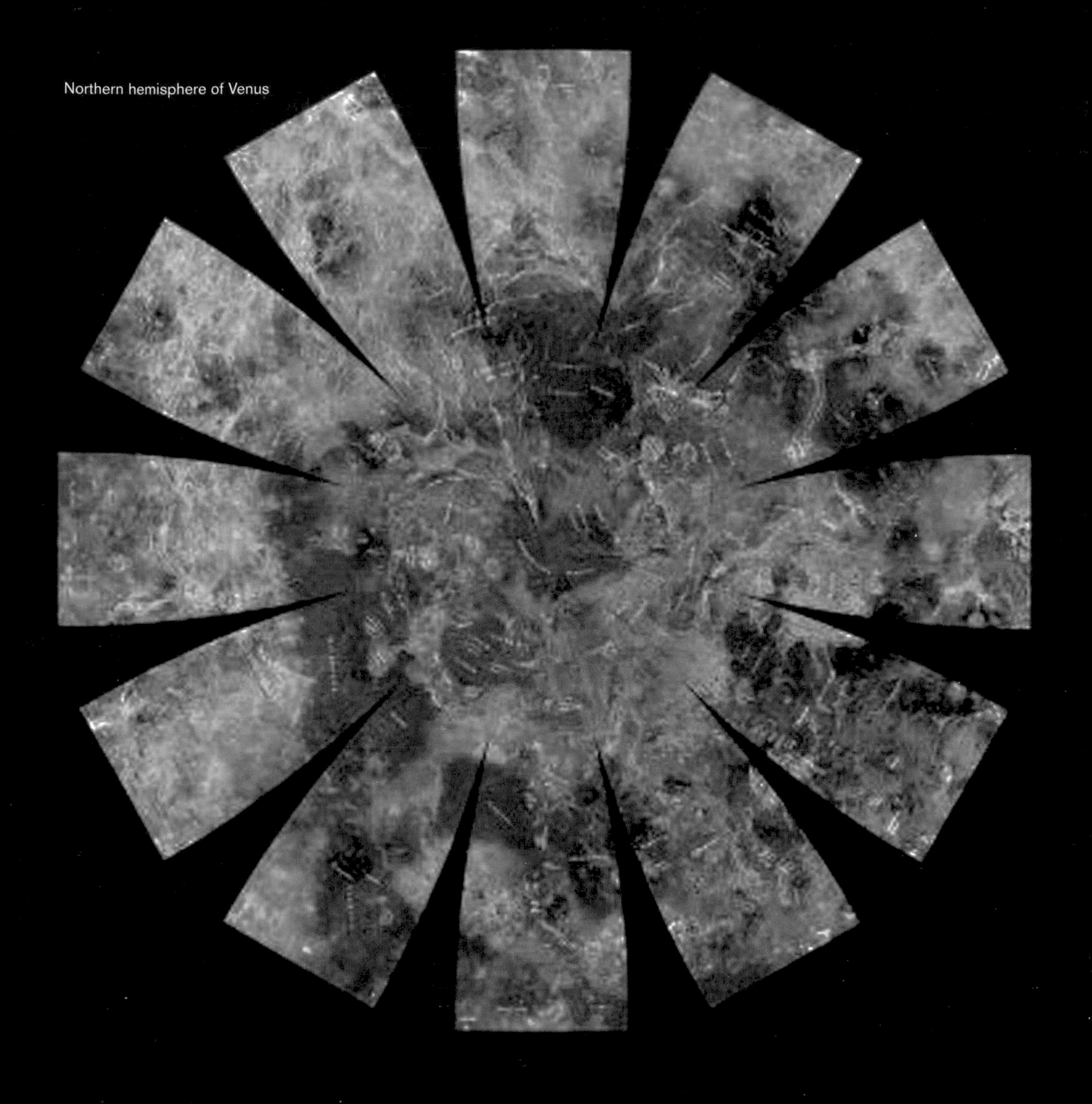
Northern hemisphere of Venus
Planetary radius (km)
6048 6050 6052 6054 6056 6058 6060 6062

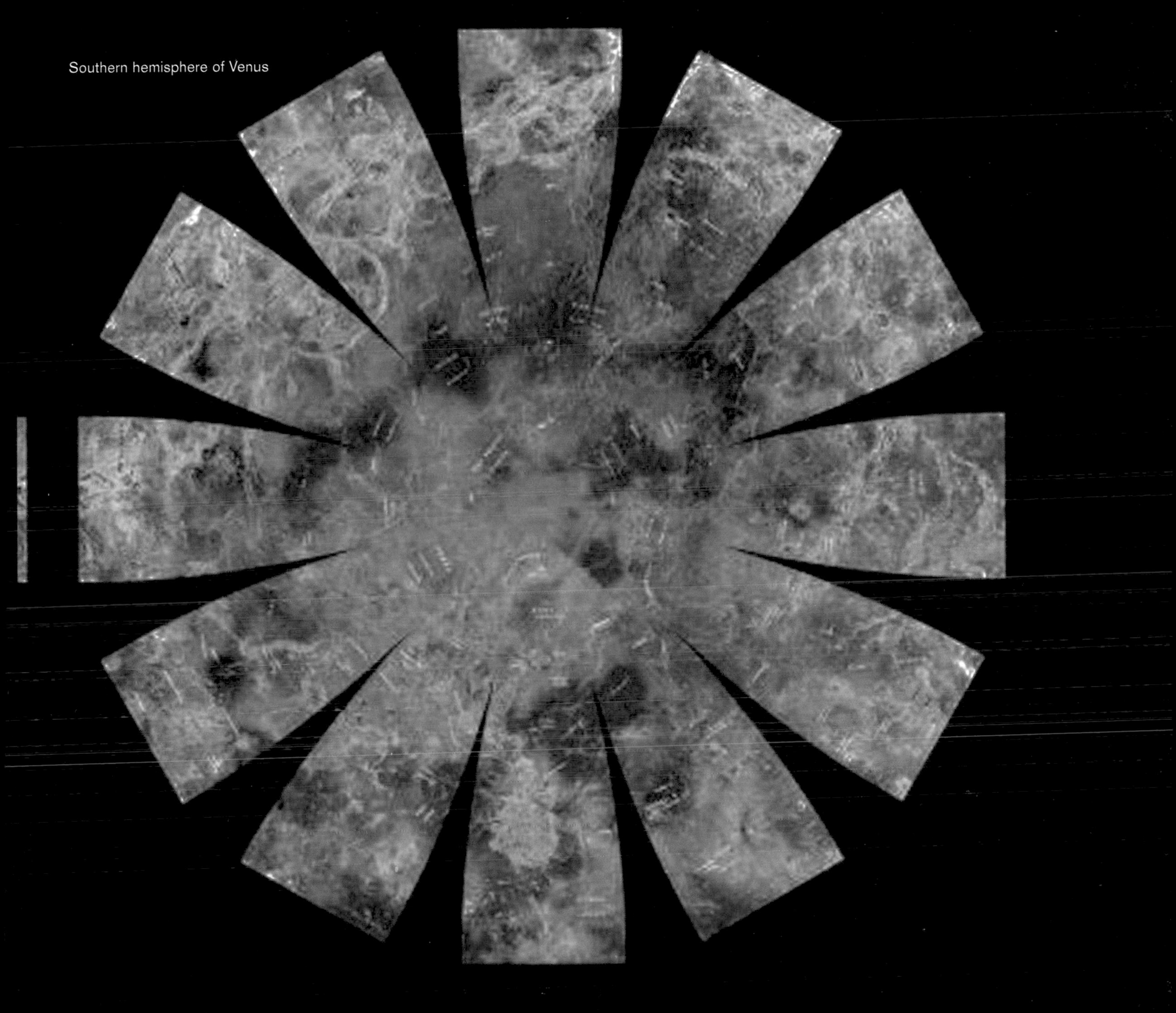

The northern and southern hemispheres of Venus revealed by more than a decade of radar investigations culminating in the 1990–94 Magellan mission. A mosaic of the Magellan images forms the image base. Gaps in the Magellan coverage were filled with images from Soviet Venera 15 and 16 spacecraft and Earth-based Arecibo radar. The composite image was processed to improve contrast and to emphasise small features and was colour-coded to represent elevation. Gaps in the elevation data from the Magellan radar altimeter were filled with altimetry from the Venera spacecraft and the US Pioneer Venus missions. NASA/JPL/USGS PIA03167

The photographs on these pages are surface photographs of Venus from the Soviet Venera 14 spacecraft. The Venera 14 lander became the second Venus surface probe to transmit color images after setting down on 5 May 1982. These pictures were taken from its two opposite facing cameras. Parts of the lander and semi-circular lens can be seen in both the left and right images. The right frame shows the lander testing arm. Data collected by Venera has supplemented data collected by Magellan to help produce computer-generated images such as the coloured hemispheric views on pages 168–69. National Space Science Data Centre.

The images on the following pages have all been generated from data collected by Magellan. As they are radar images, they need to be interpreted differently to images made in ordinary light. Generally, bright areas indicate a rough surface, while dark areas indicate a smooth one.

Pages 172–73 This Magellan full resolution mosaic shows an area 160 kilometres by 250 kilometres in the Eistla region of Venus. The prominent circular features are volcanic domes, 65 kilometres in diameter. Sometimes referred to as 'pancake' domes, they represent a unique category of volcanic extrusions on Venus formed from viscous (sticky) lava. The cracks and pits commonly found in these features result from cooling and the withdrawal of lava. A less viscous flow was emitted from the northeastern dome towards the other large dome in the southwest corner of the image. Venus–Eistla Region, NASA PIA00084

Page 174 Three large meteorite impact craters, with diameters that range from 37 to 50 kilometres, can be seen in this image of the Lavinia region of Venus. Situated in a region of fractured plains, the craters show many features typical of meteorite impact craters, including rough (bright) material around the rim, terraced inner walls and central peaks. Numerous domes, probably caused by volcanic activity, are seen in the southeastern corner of the mosaic. The domes range in diameter from 1 to 12 kilometres. Some of the domes have central pits that are typical of some types of volcanoes. North is at the top of the image. Venus–Lavinia Region Impact Craters, NASA PIA00214

Page 175 This view of the

surface of Venus acquired by the Magellan spacecraft shows a geographically young region of lowland plains. The location is near the equator between two highland areas known as Asteria Regio and Phoebe Regio. Complex canyon systems that trend northeast and northwest were produced as Venus' crust was pulled apart by extensional forces. Some were filled with younger lava flows. The canyons are typically 5 to 10 kilometres wide, 50 to 100 kilometres long and rimmed by fault scarps 100 metres or so high. Venus–Asteria Regio and Phoebe Regio, NASA PIA00237

Page 176 A 100-kilometre-wide nova superimposed on Yavine Corona, a 500-kilometre-wide asymmetric feature. Coronae are roughly circular, volcanic features believed to form over hot upwellings of magma within the Venusian mantle. This is a three-dimensional perspective view of Venusian terrains composed of reduced resolution left-looking synthetic-aperture radar images merged with altimetry data from the Magellan spacecraft. Nova superimposed on Yavine Corona, NASA PIA00150

Page 177 This Magellan image covers part of the eastern flank of the volcano Sapas Mons on the western edge of Atla Regio. The bright lobate features along the southern and the western part of the image are lava flows. These flows range in width from 5 kilometres to 25 kilometres with lengths of 50 kilometres to 100 kilometres, extending off the area shown here. Located near the centre of the image is a 20-kilometre diameter impact crater. This crater is superimposed on a northeast/southwest trending fracture while the southern part of the crater's ejecta blanket is covered by a 6-kilometre-wide radar-bright lava flow. This is one of only a few places on Venus in which an impact crater is seen to be covered by volcanic deposits. East Part of Sapas Mons with Flooded Crater, NASA PIA00099

Pages 178–79 This Magellan radar image is of a 'half crater' located in the rift between Rhea and Theia Montes in Beta Regio on Venus. The 37-kilometre wide crater was initially named the Somerville crater, but is now called after the American Nobel Prize winner Emily Balch. It has been cut by many fractures or faults since it was formed by the impact of a large asteroid. Fractured Somerville crater in Beta Regio, NASA PIA00100

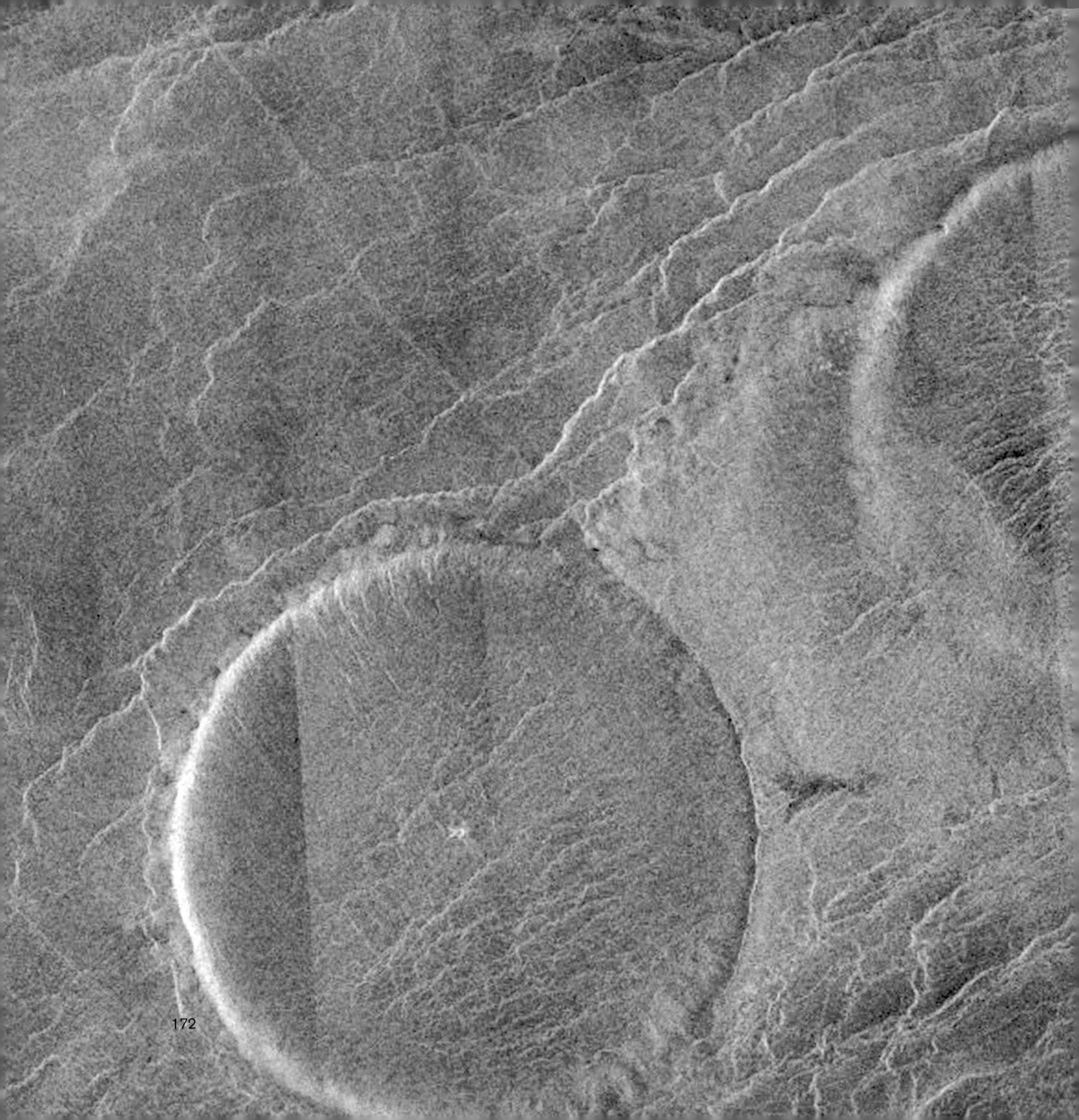

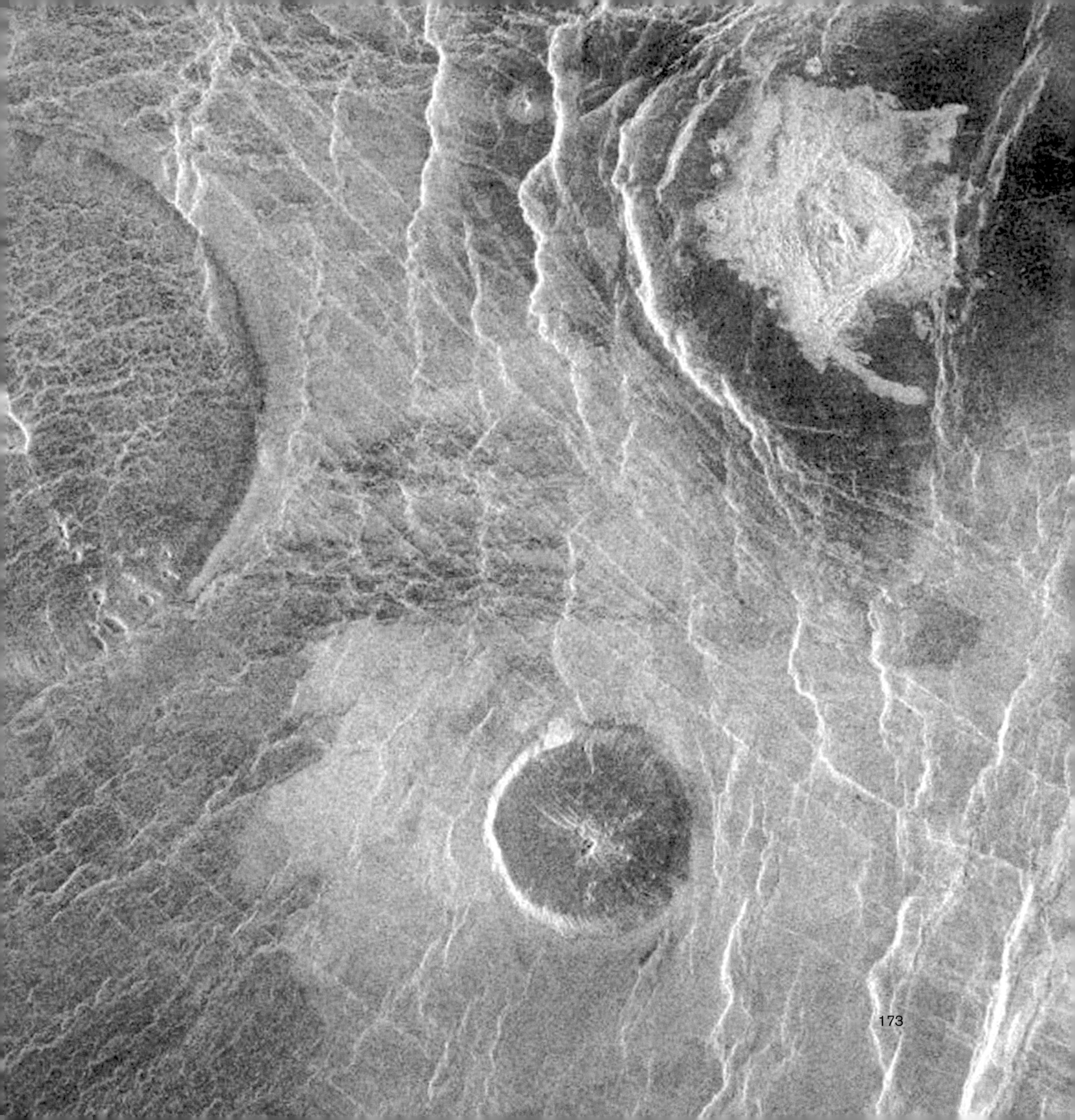

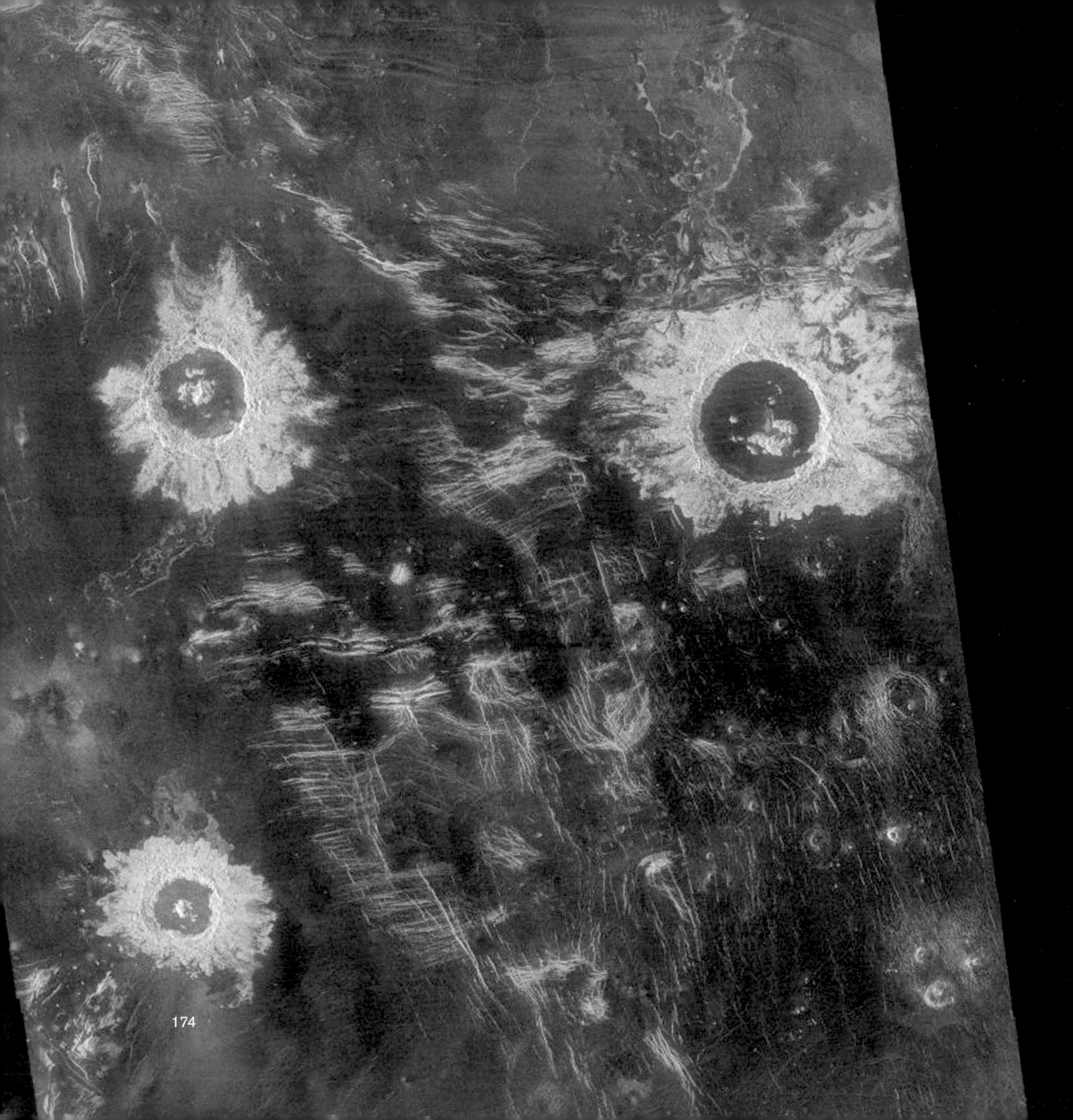

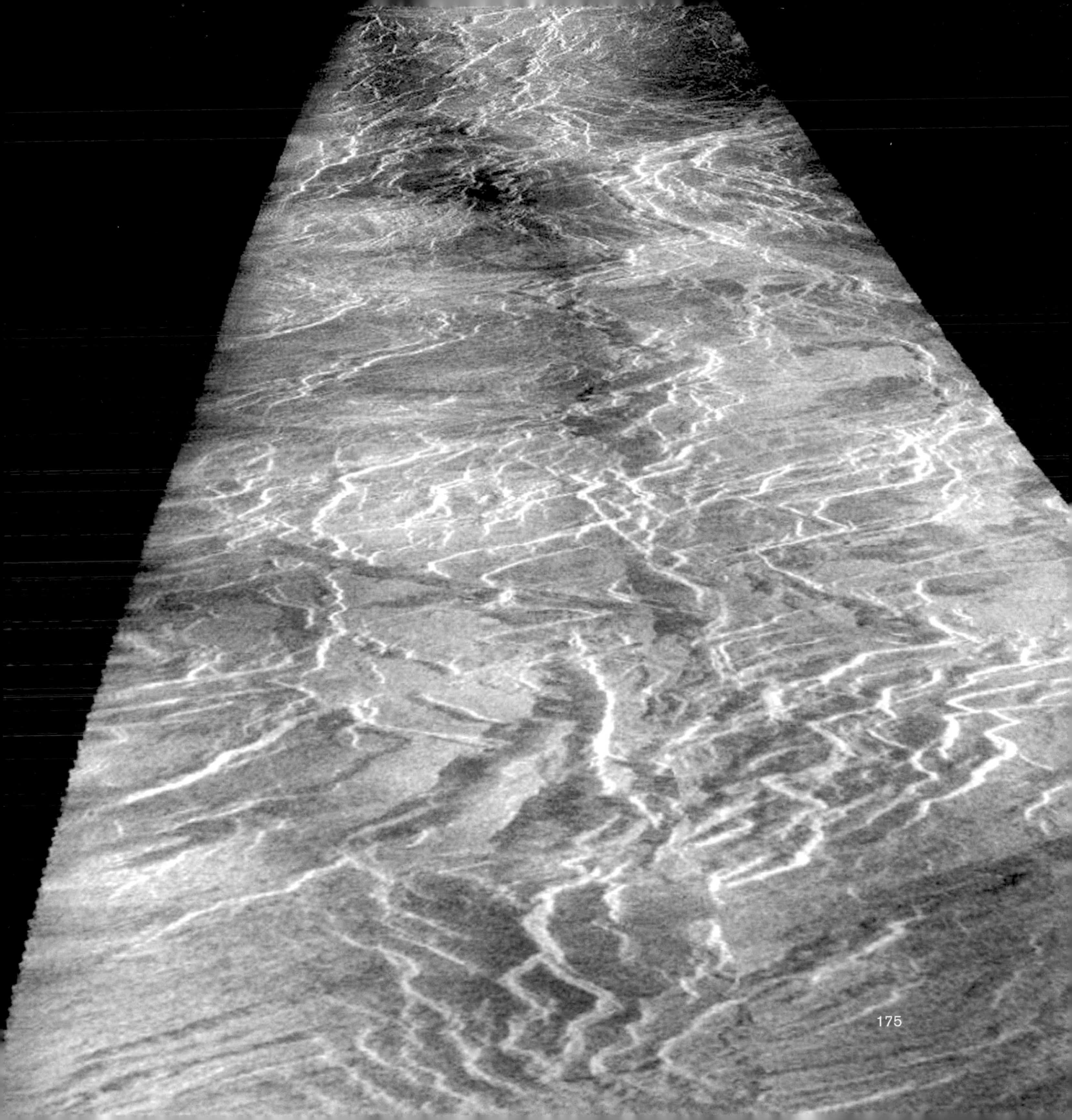

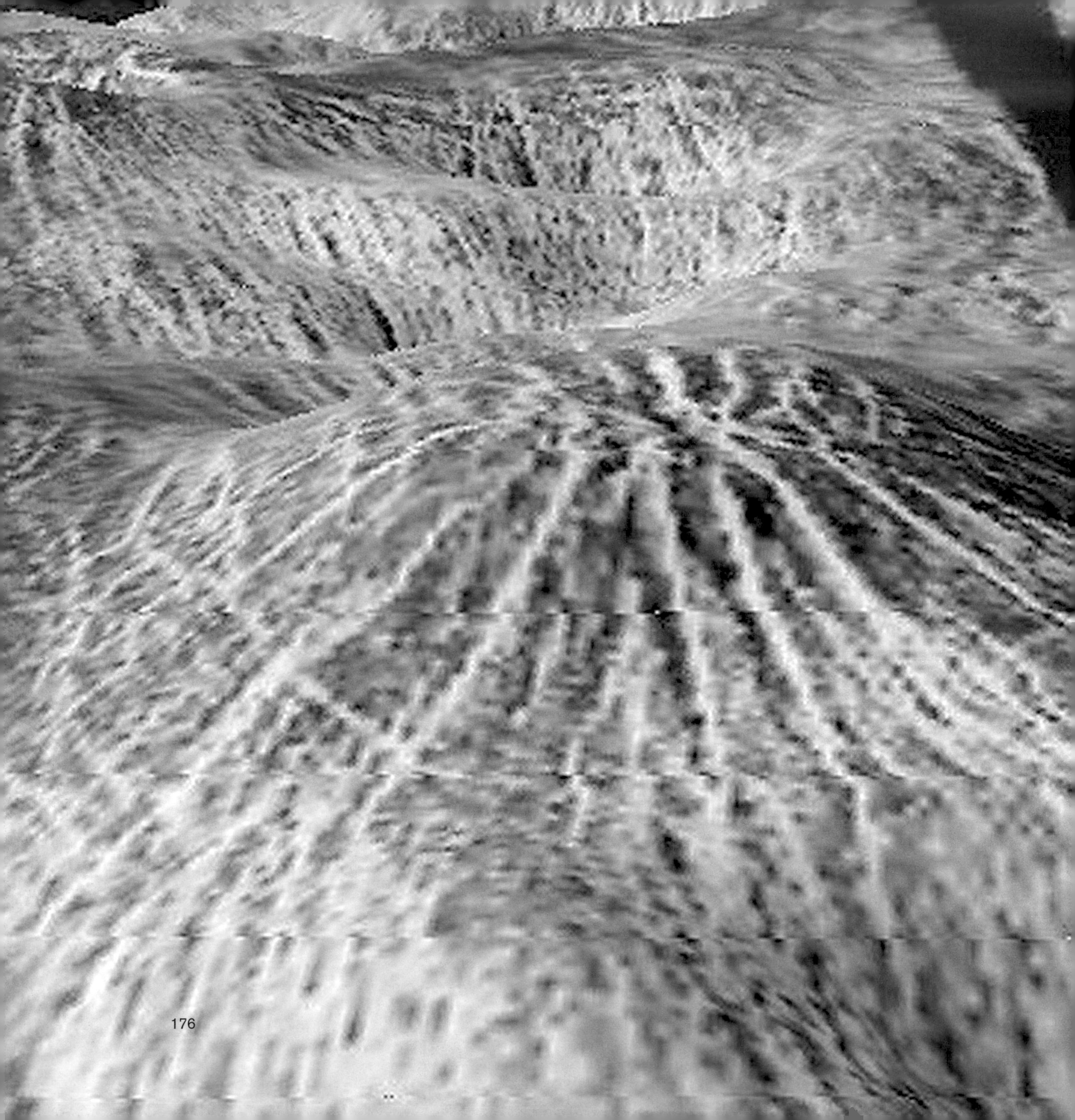

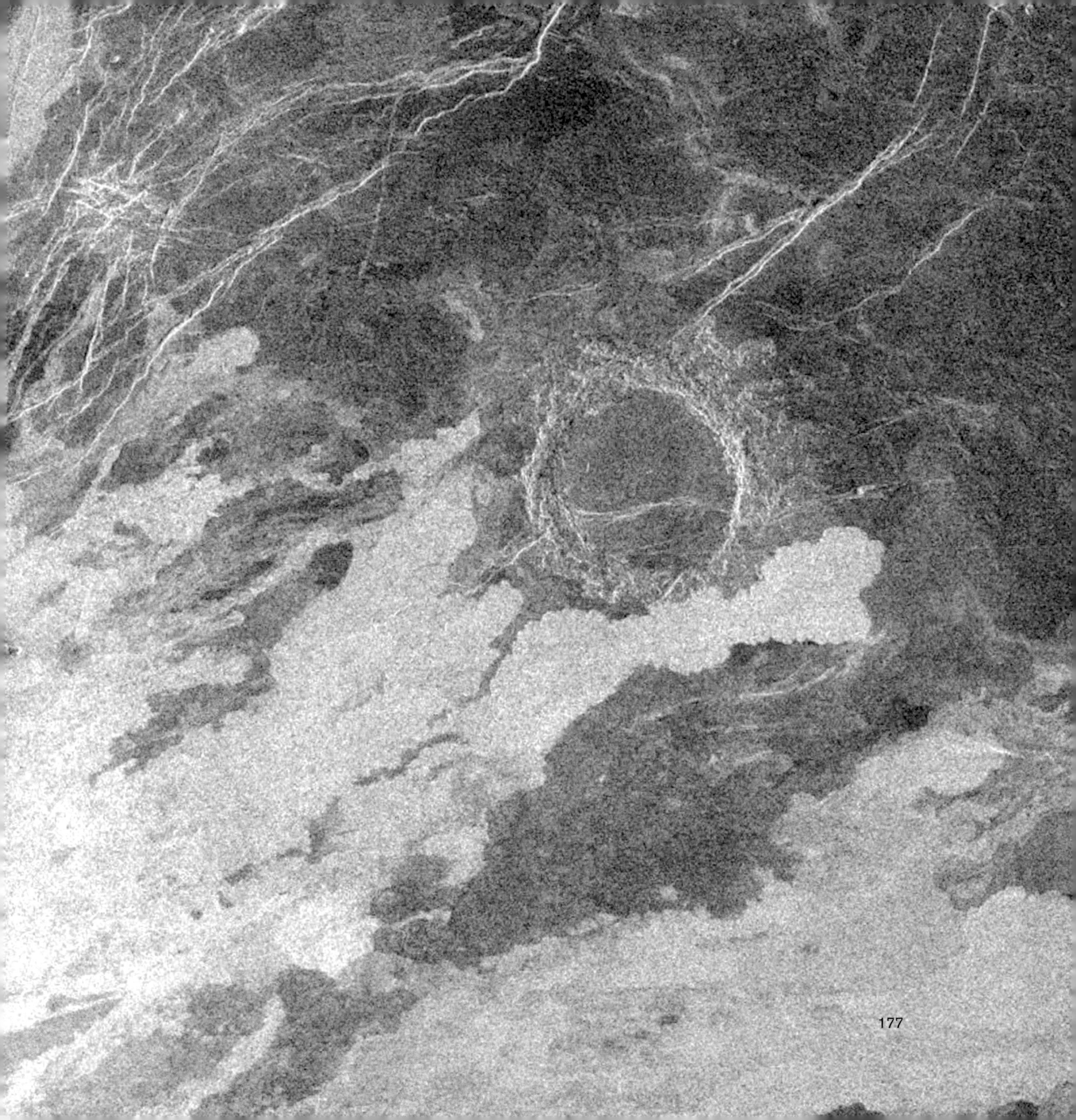

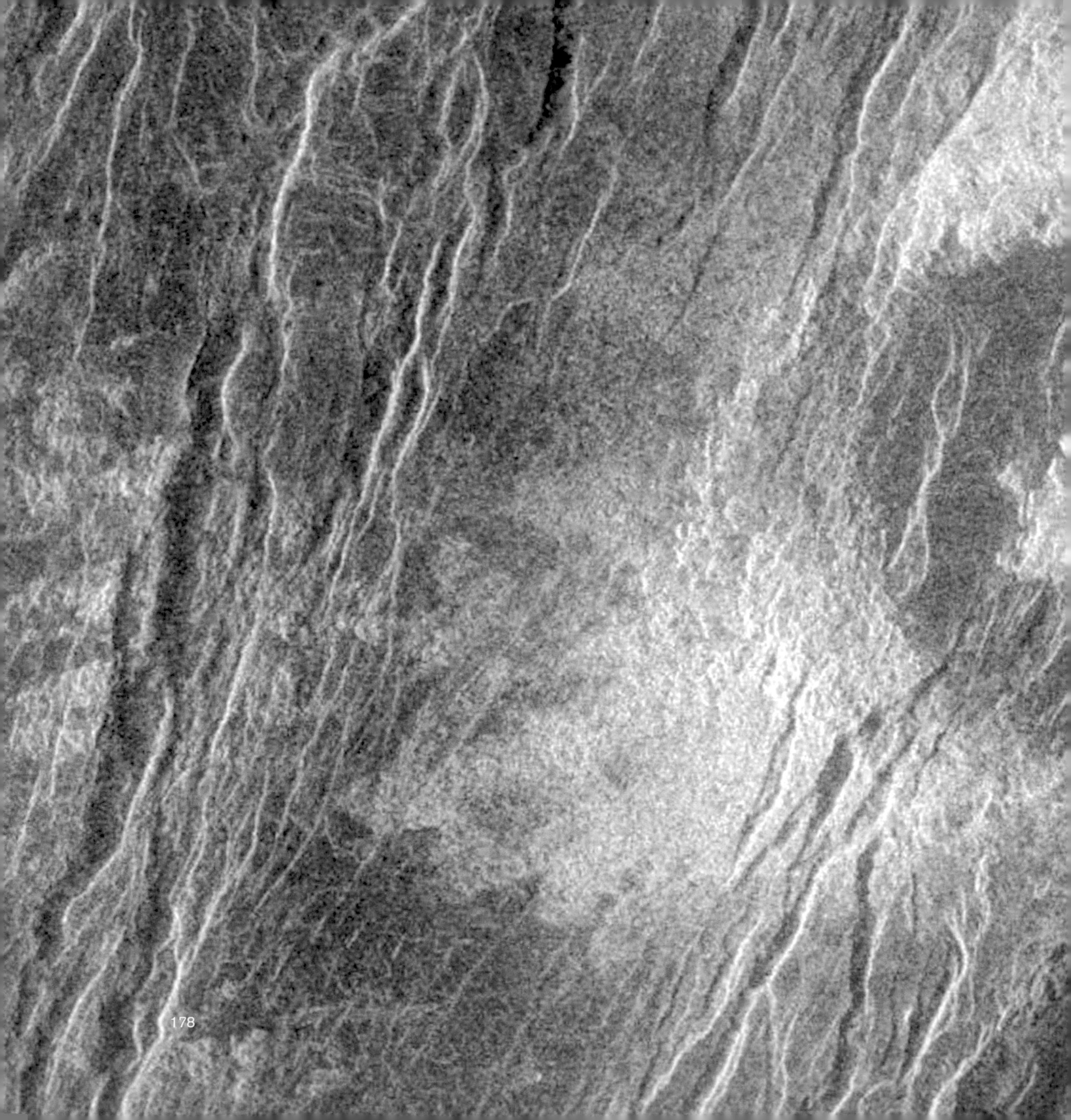

'We have been waiting for 122 years to see a transit of Venus, but our wait is almost at an end!' That was how I began a talk to several hundred people assembled in the marquee at Sydney Observatory on the afternoon of Tuesday, 8 June 2004. It was a glorious afternoon with the sky completely clear and, for early winter, the weather was a pleasantly warm 22°C, five degrees above the average maximum for the day.

The 2004 transit was quite fortuitous in that most of the globe could see at least part of the event. Unlucky places that missed out completely included New Zealand, parts of the west coast of the United States and the southern parts of South America. However, for most of Asia, Europe and most of Africa, the transit could be seen from beginning to end. For the western parts of Africa, the eastern parts of the United States and the eastern parts of South America, the transit was already in progress as the Sun rose and, where the weather permitted, the end of the event or egress could be seen.

Sydney Observatory had prepared as carefully for the 2004 transit as it had for the 1874 and 1882 transit observations. The difference was that in the 1800s the preparations were to carry out serious scientific research during the transit, while in the 21st century the emphasis was on informing the public about the importance and history of the transit and helping people to observe the event. Observing the event with modern telescopes, modern eyes and modern equipment provided an opportunity to understand the complex descriptions of the event left behind by the likes of Captain Cook and Henry Chamberlain Russell and to appreciate the difficulties that they faced in making their observations and timings.

With my talk at an end, it was time for the transit. Visitors lined up to view the sun through a line of telescopes set up on the western side of the Observatory grounds. Each telescope had a special filter placed at the far end of the tube to block out most of the sunlight. Not available to observers at previous transits, such filters are much safer than the now banned filters placed near the eyepiece. A staff member looked after each telescope to assist visitors.

I began a live radio interview just as first contact was due to happen. Pressing my way to the head of one of the queues, past a line of expectant but good-natured people, I described to the radio audience what I could see through the telescope. What I saw was hardly spectacular – a slight indentation on the Sun's edge – but I was witnessing the first time Venus had crept in front of the Sun in the lifetime of anyone alive.

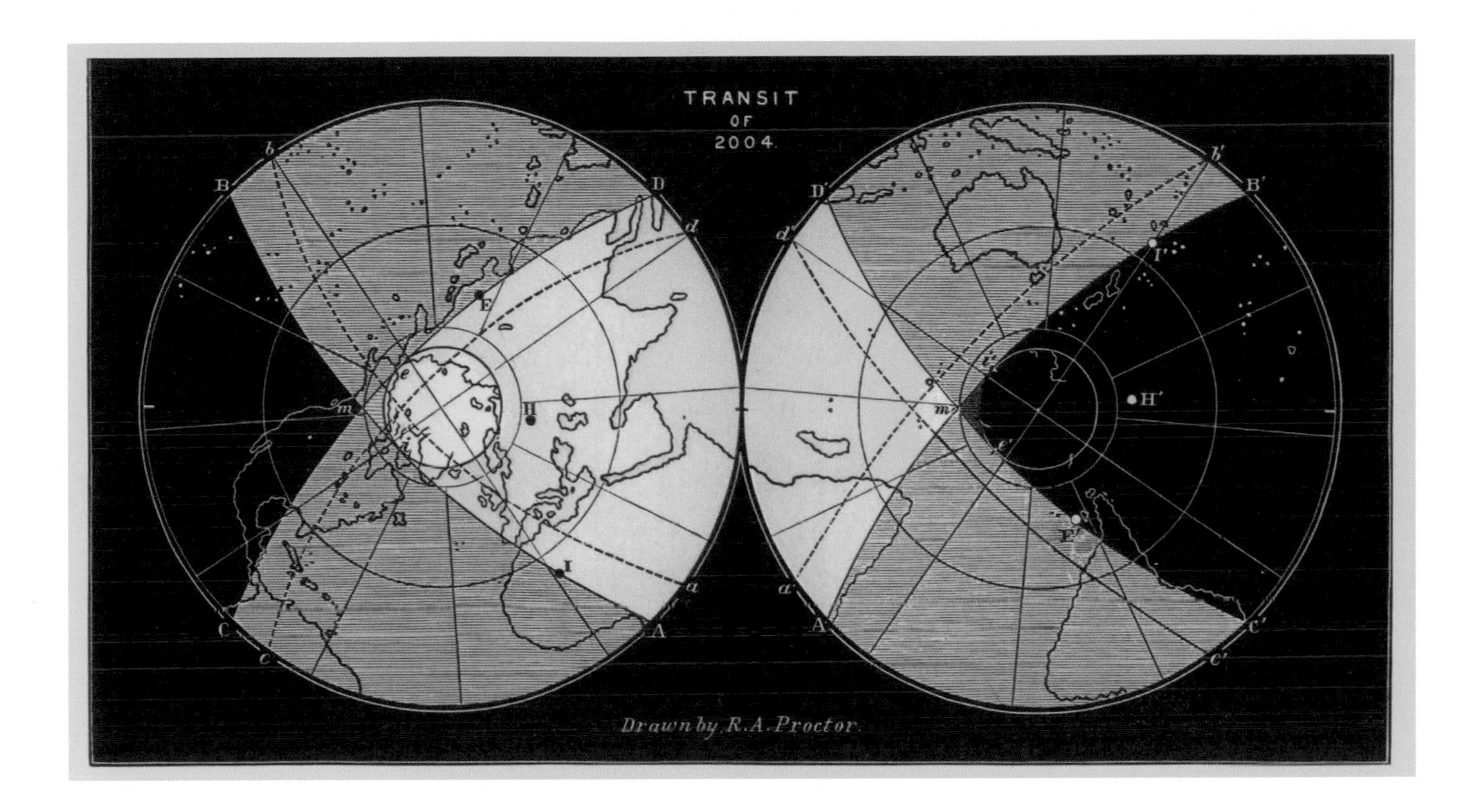

The visibility of the 2004 transit of Venus, with the northern hemisphere on the left and the southern hemisphere on the right. The brightest area denotes the regions from which the transit was fully visible, the shaded areas indicate regions from which the beginning or end of the transit was visible, and the darkest areas are the regions from which the transit could not be seen. Map from *Transits of Venus* by RA Proctor, 1874. Brian Greig collection

As the transit progressed, Venus was seen as a clearly visible dark spot on the face of the Sun. A 17-year-old visitor seemed more impressed with the queues at the telescopes than the phenomenon in progress, telling the *Sydney Morning Herald*, 'It's a pretty big turnout for a dot'. Others were more entranced, such as one visitor who said, 'It's a whole planet, but it looks so small compared to the sun. It just shows you the phenomenal size of the solar system'. An imaginative seven-year-old girl said, 'It was like a little hole in the side of the sun'.

The south dome of the Observatory contains the 11½-inch (29-centimetre) telescope that was made

The historic 11½-inch (29-centimetre) telescope in the south dome of Sydney Observatory has now been used to observe both the 1874 and 2004 transits of Venus. Collection: Powerhouse Museum, Sydney. Photo: Chris Brothers

by Hugo Schroeder of Hamburg, Germany, and was installed by Russell to observe the 1874 transit. On this occasion for safety reasons the large lens was not used. Instead visitors could see Venus moving across the face of the Sun projected from the smaller 'finder' telescope on to a white screen. When the tracking mechanism that allows the telescope to follow the motion of the Sun broke down, the conservator who was looking after the telescope at that time, Keith Potter, stepped in and wound the telescope by hand. Keith considered that the telescope was letting him know that it had already seen a transit of Venus.

A modern computer-controlled telescope stands in the north dome of the Observatory, the dome that Russell had built five years after the 1874 transit. Using this telescope, staff members made observations with a hydrogen alpha filter, which only lets through red light emitted by hydrogen atoms. The Observatory's senior astronomy educator, Geoff Wyatt, took a series of images of the transit in that red light. One image clearly showed that the black drop effect is real and can be photographed even through a modern telescope (see page 165). It won first prize in an astrophotography competition, the David Malin Awards judged by the world's most famous astrophotographer, Dr David Malin.

When the Sun, with Venus still part of the way across, set an hour and three-quarters after the beginning of the transit, no one was disappointed at Sydney Observatory. Visitors and astronomers alike had seen enough of one of the great events of astronomy to be satisfied.

MASS TRANSIT

Many people watched the transit with improvised equipment from their workplaces or schools. As long as they were using the right filters or, preferably, projecting the images through a small telescope or a pair of binoculars, they were safe and not damaging their eyes.

Elsewhere in Australia, observatories and local amateur astronomical groups organised their own events, some of which took place at locations used for the 1874 transit, including Woodford in the Blue Mountains, one of Russell's observing sites, Campbell Town in Tasmania, where Captain Raymond's party had observed, and Melbourne Observatory, from where Ellery had made his observations. In Melbourne, the Astronomical Society of Victoria organised a viewing session on top of a shopping centre car park in the inner suburb of Richmond. In the wine-growing district of the Hunter Valley in New South Wales, a Venus Transit Party took place at one of the wineries.

People watched the transit from numerous places around the world. A group of 300 British enthusiasts, fearing clouds in the UK, flew to the Red Sea resort of Sharm el Sheikh, near the tip of the Sinai Peninsula in Egypt. One of the participants, Nick James, relates in the *Journal of the British Astronomical Association* that he did not see a single cloud during the four-day stay. He had his telescope and other equipment set up near the hotel's roof-top bar. From there he saw first contact take place, unlike the BBC which set up equipment on an adjacent roof and managed to concentrate on the wrong side of the Sun and missed Venus just touching the Sun.

James records that around the time of the second contact he saw the black drop effect for almost a minute, 'but I am convinced that it is a seeing effect and nothing more'. By 'seeing', he means a technical term denoting the amount of blurring of the image of the Sun by the Earth's atmosphere. Apparently, the air currents created by the hot concrete roof made seeing particularly bad. It was also extremely hot – 45°C in the shade – and James decided to cool off in the pool, from where he continued watching the Sun using a solar filter.

Others took the risk of poor weather and stayed in the UK. The famous British astronomer, author and television presenter Sir Patrick Moore invited an elite group to observe the transit from his garden at Selsey in Sussex, southern England. To the pleasurable surprise of the participants, the sky was clear from the beginning to the end of the transit. Astronomy author Martin Mobberley relates in the *Journal of the British Astronomical Association* that among all the high-technology equipment, Sir Patrick observed the transit by projecting an image of the Sun through the 3-inch (75-millimetre) brass-tubed lens telescope that he had bought as a child in the 1930s for the sum of £7 and 10 shillings. Participants were regularly asked to keep quiet as Sir Patrick was put live to air on the BBC.

Mobberley states that those using smaller telescopes and those projecting the image saw the black drop. However, in 'the rarest sharp moments, when the atmosphere was still, there was no black drop at all, just a sharp black disk with a clear channel of Sun to the solar edge … fascinating'. Sir Patrick summed up the event by saying, 'It can hardly be called spectacular, but it is exceptionally rare: no-one alive has seen one of these before today'.

In the United States the transit could only be seen in the eastern part where the egress or end of the event could be observed in the early morning after sunrise. New York City amateur astronomers

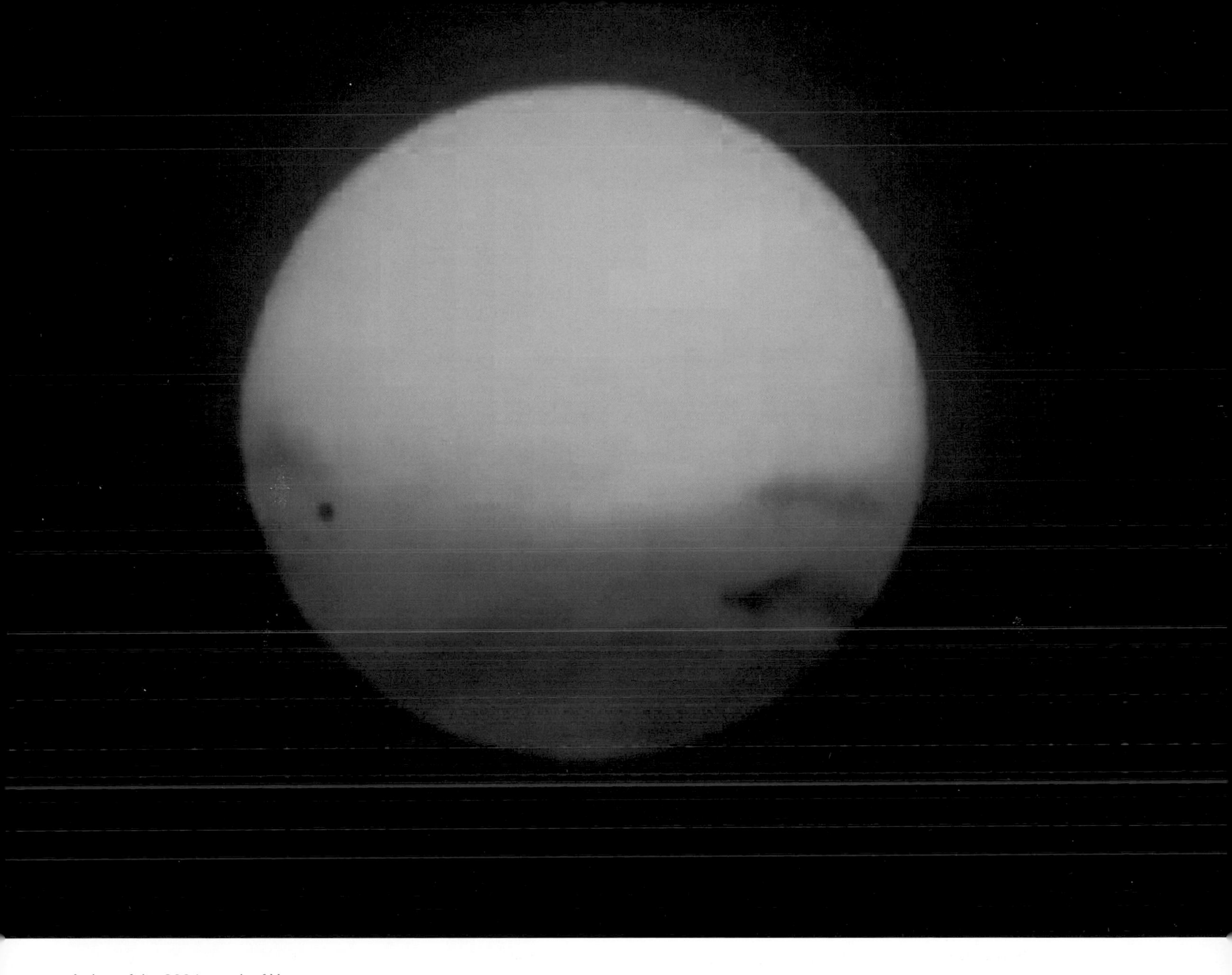

A view of the 2004 transit of Venus from Carl Schurz Park, Manhattan, New York. Photo: Rik Davis of the Amateur Astronomers Association of New York

Astronomy author Sir Patrick Moore (seated) watched the 2004 transit of Venus from the garden of his home at Selsey, Sussex, England, surrounded by friends, astronomers and television crews. Courtesy Martin Mobberley and the *Journal of the British Astronomical Association*

> Astronomers observing the 2004 transit of Venus by the swimming pool in Sharm el Sheikh, Egypt. Courtesy Hazel McGee and the *Journal of the British Astronomical Association*

Sally Russell, a research chemist, artist and keen amateur astronomer, sketched the ingress on 8 June 2004 from the courtyard of the Royal Observatory, Greenwich, during a free, public event attended by hundreds of people. Her observations were made through a 4-inch (10-centimetre) Astrophysics lens telescope and a white light filter while assisting a long queue of people wanting to look through her telescope. The times of the observations were, from left to right, 05.30, 05.38, 05.40 and 05.43 UT (Universal Time). Courtesy Sally Russell

A NEW BREED OF SOLAR TELESCOPE

The world's highest-resolution solar telescope obtained remarkably clear images of the 2004 transit.

ALSO WATCHING THE TRANSIT was the world's highest resolution solar telescope, the Swedish 1-metre Solar Telescope (SST), which at that stage had been operating for just two years. The SST has an excellent location, a high mountaintop on the Spanish island of La Palma in the Canary Islands, where a dozen or more research telescopes share the space.

Telescopes observing the Sun are necessarily in the sunlight, and the Sun's heating effects on the optics and on the air inside the equipment act to blur the image. The SST minimises the effect of this heating by using a vacuum inside the telescope tube, with the telescope's 1-metre wide lens used to seal off the vacuum. As the lens focuses 700 watts of heat from the Sun into the telescope there is a lot of heat generated. To further reduce the effect of this heat, appropriate parts of the telescope optics are cooled by circulating a mixture of water and glycol through them.

Unusually for astronomy, the lens is made of a single piece of glass. A single piece of glass, however, disperses the light into a spectrum and does not bring different colours together to form a sharp image. To overcome this, telescope lenses are normally made of two or more pieces of glass with different optical properties that are carefully arranged to bring at least two colours to the one focus. Instead, the SST employs an old and rarely used design published by German architect

Three images of Venus as it left the Sun at the end of the 2004 transit made by the Swedish 1-metre Solar Telescope. The original monochrome images have been processed to provide colour. The times of the images, from top to bottom, expressed in Universal Time (UT) are 11.07.09 (11 hours 7 minutes 9 seconds), 11.12.06 and 11.15.42. Courtesy Dan Kiselman and the Institute for Solar Physics of the Royal Swedish Academy of Sciences

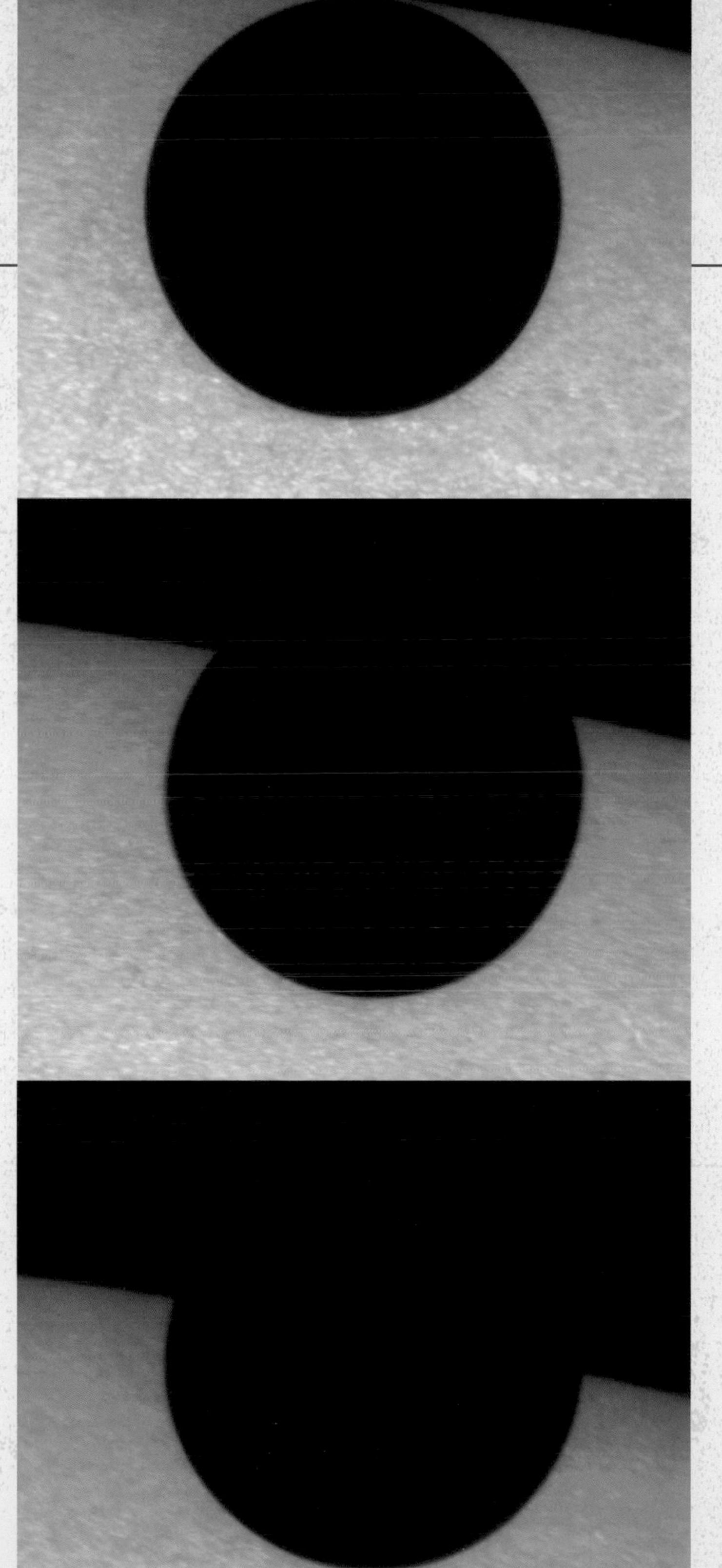

Ludwig Schupmann in 1899. In this design the beam of light from the main lens is directed towards a concave or negative lens with a flat mirror behind it. The beam passes twice through the corrector lens with the result that colour dispersion is removed and all colours are brought to the one sharp focus. The SST is the first large telescope that successfully uses this system.

Another very modern feature of the telescope is that it uses adaptive optics. This technique measures the instantaneous blurring in the image due to the atmosphere and then deforms a specially designed mirror inside the instrument to compensate. By doing this accurately and at a rate of hundreds or maybe a thousand times a second, much clearer images can be obtained.

With this sophisticated design, the SST obtained remarkably clear images of the 2004 transit. Henry Chamberlain Russell, Robert LJ Ellery and other observers of the 1874 and previous transits would have been impressed and envious.

gravitated towards Carl Schurz Park, which is a peaceful park located on top of the six-lane FDR Drive with a clear view to the northeast horizon over the East River. For the transit there were about 30 astronomers present with their telescopes, about 300 public and a television broadcast van. Although haze covered the Sun as it rose, good views were obtained from about 6 am EDT (Eastern Daylight Time), despite some shaking of the telescopes by the heavy traffic below. Prominent New York amateur astronomer John Pazmino relates that for the next hour until Venus left the Sun people took leisurely looks through the telescopes and, 'There was time to munch a bagel, sip coffee, tend to babies, exercise dogs, read newspapers'.

OBSERVING FROM SPACE

On 8 June 2004 the transit of Venus was not only watched from places around the globe, it was for the first time also observed from space. NASA's Transition Region and Coronal Explorer (TRACE) spacecraft made a series of high-resolution images. In a scientific paper published in *The Astronomical Journal* in 2011 two astronomers from the United States, Jay M Pasachoff and Glenn Schneider, together with an astronomer from the Observatoire de Paris, Thomas Widemann, reported on the observations.

Although the spacecraft is outside Earth's atmosphere, a slight black drop effect can be seen at the internal contacts. The observation reinforces the conclusion that the black drop can be easily explained. As part of the explanation they note that the edge of the Sun is darker than the rest of the disc: the Sun is a ball of gas and at the edge a telescope is looking at higher and hence cooler and darker levels. Astronomers call this phenomenon 'limb darkening'.

Pasachoff and Schneider's explanation of the black drop is that it comes from a combination of limb darkening with the spreading of light by the atmosphere and the telescope. On this basis the black drop is more likely to be seen with a small aperture telescope because that spreads the light to a greater extent than one with a larger aperture. Also, it is more likely to be seen if the atmospheric conditions are poor or if the Sun is seen low in the sky at sunrise and sunset. These predictions seem to well match the observational reports.

Pasaschoff and Schneider concentrate on the narrow ring of light (aureole) visible in the part of Venus off the Sun as it heads towards second contact or after the third contact. As proposed by the Russian scientist Mikhail Lomonosov from his observations

For the first time ever, a transit of Venus was observed from space, thanks to NASA's TRACE satellite.

TRACE

During more than 12 years of successful operation, the Transition Region and Coronal Explorer (TRACE) satellite produced millions of images, including a series of images of the transit of Venus in 2004. TRACE was a mission of the Sanford-Lockheed Institute for Space Research and part of the NASA Small Explorer Program.

Around ingress, when the disc of Venus enters the sun's disc, and around egress, when Venus leaves the sun, two interesting phenomena may occur: the elusive aureole effect and the notorious black drop effect (see box, 'The Black Drop', page 98). These phenomena have been reported by observers ever since the 1761 transit, but their origin was only properly understood after scientists studied computer models and high-resolution images from the TRACE satellite.

TRACE was officially retired by NASA in 2010 and was replaced by the Solar Dynamics Observatory (see page 215).

This image shows Venus on the eastern limb of the Sun. The faint ring around the planet comes from the scattering of its atmosphere, which allows some sunlight to show around the edge of the otherwise dark planetary disc. The faint glow on the disc is an effect of the TRACE telescope. NASA/LMSAL

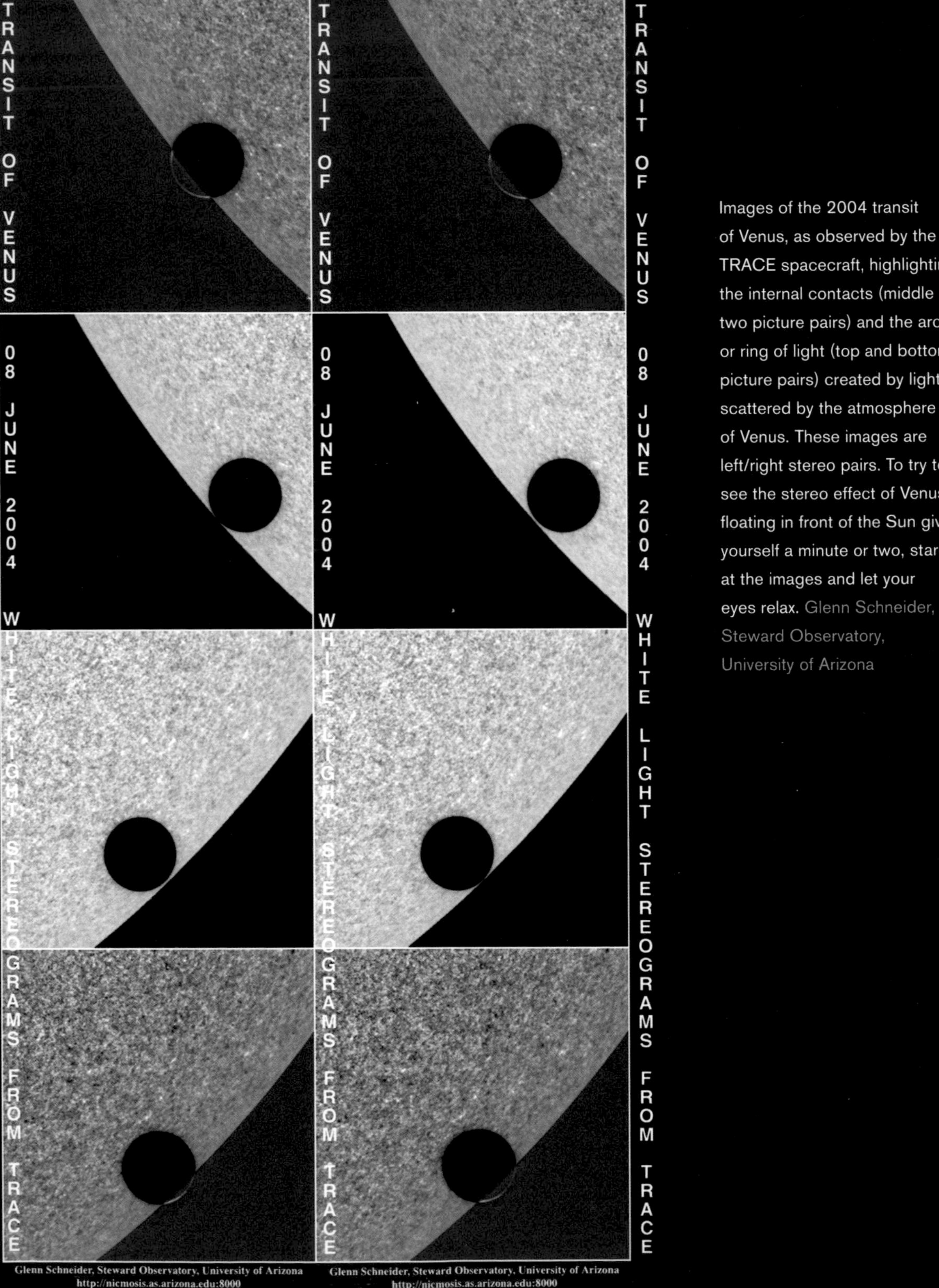

Images of the 2004 transit of Venus, as observed by the TRACE spacecraft, highlighting the internal contacts (middle two picture pairs) and the arc or ring of light (top and bottom picture pairs) created by light scattered by the atmosphere of Venus. These images are left/right stereo pairs. To try to see the stereo effect of Venus floating in front of the Sun give yourself a minute or two, stare at the images and let your eyes relax. Glenn Schneider, Steward Observatory, University of Arizona

Image of Venus transiting the sun in 2004, captured by the TRACE spacecraft in extreme ultraviolet. NASA/ LMSAL

at St Petersburg during the 1761 transit, this ring of light is from Venus' backlit atmosphere bending sunlight towards the Earth. What is interesting is that there is a definite asymmetry involved, so that parts of the planet's atmosphere appear brighter than others. The authors remark that Henry Chamberlain Russell noted a similar asymmetry, as for example at his observations at egress (see page 148).

It seems that this asymmetry can be explained with the help of previous detailed observations of Venus' atmosphere made by spacecraft circling the planet. The spacecraft have found that there are rings of cooler air surrounding the north and south poles of the planet. The cooler air pushes down the height of the cloud tops above the surface by about 8 to 10 kilometres. With the cloud tops lower, more sunlight is bent towards the Earth and we see the polar regions of the atmosphere as brighter. During the 1874 transit, Russell saw a brighter 'polar spot' in the northern hemisphere of the planet, while in 2004 TRACE saw a similar effect in the vicinity of the planet's south pole.

Sapas Mons is a volcano on Venus that is 400 kilometres across and 1.5 kilometres high. The volcano is displayed in the centre of this computer-generated three-dimensional perspective view of the surface of Venus. Lava flows extend for hundreds of kilometres across the fractured plains shown in the foreground. Venus – 3D Perspective View of Sapas Mons NASA/JPL PIA00107

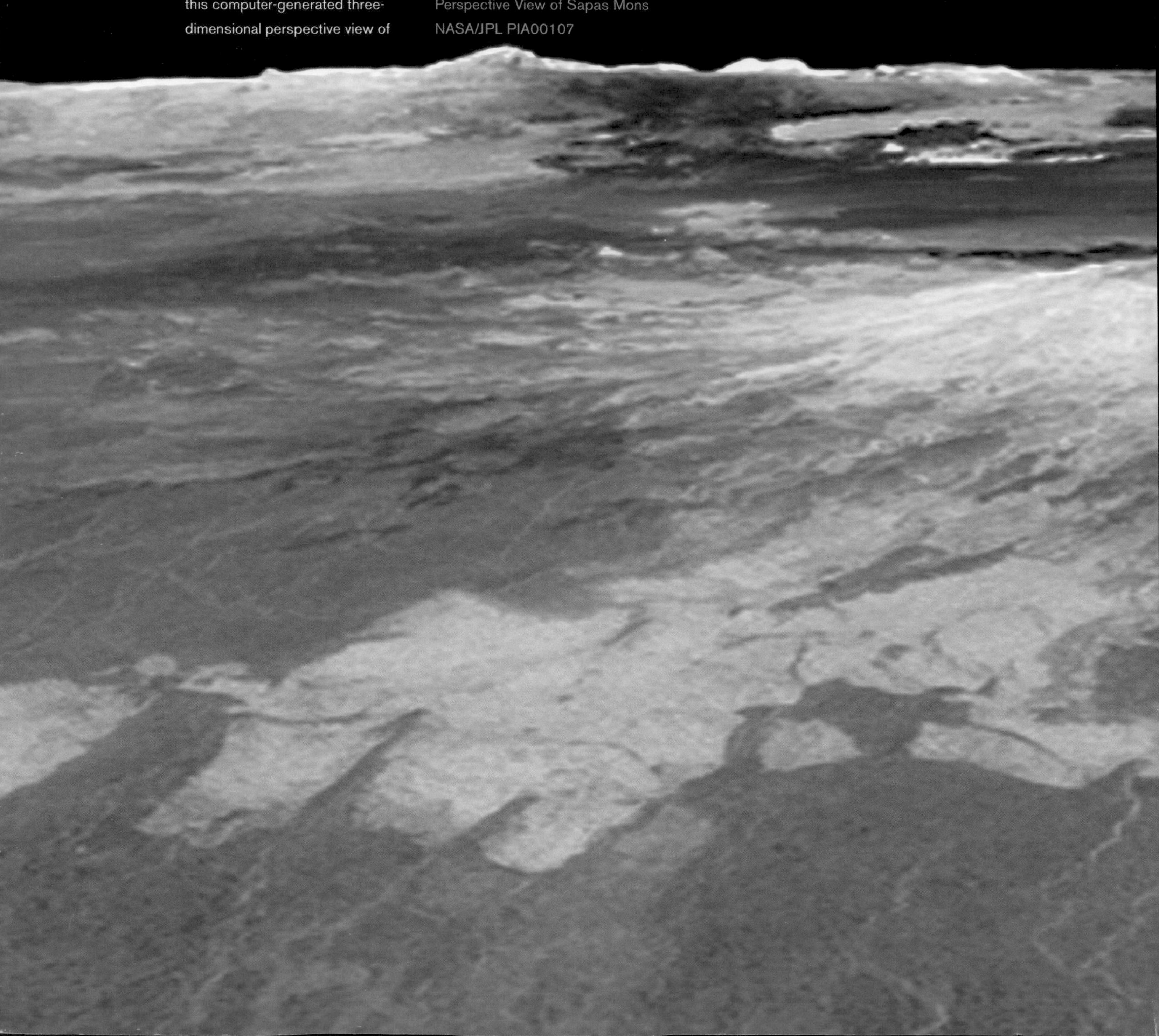

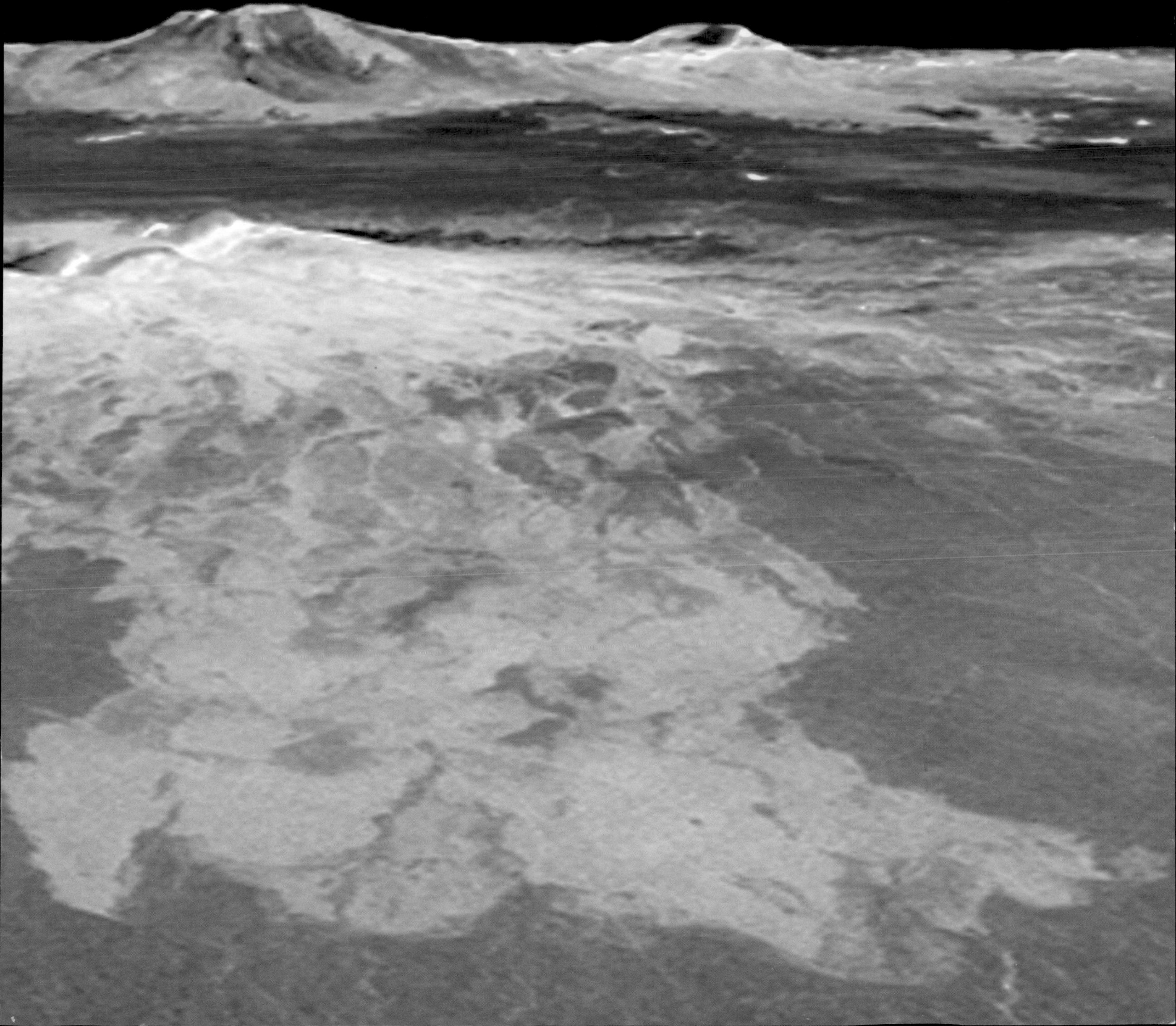

USING THE TRANSIT METHOD TO FIND OTHER EARTHS

A regularly repeating dip in the brightness of a star indicates to scientists on the ground the possibility of a planet circling that star.

ANOTHER SPACE-BASED telescope constantly searches for transits of planets. NASA's Kepler spacecraft, launched in March 2009, is not seeking the transits of Mercury or Venus that astronomers have observed over the last few centuries from Earth; instead, it is searching for the transits of planets across the faces of distant stars. Even from space, stars appear only as pinpoints of light, so that a transit is detected as a temporary slight dimming in a star's brightness.

NASA placed the spacecraft in an independent path around the Sun with a period exceeding that of Earth's year by six days. As a result, the spacecraft is slowly drifting away from the Earth. This arrangement ensures that Kepler's view is never impeded by the Earth or the Moon.

The instrument on board the 1-tonne spacecraft is a 0.95-metre (37-inch) telescope that constantly points towards an area in the vicinity of the northern hemisphere constellations Cygnus and Lyra. In that area, the telescope is constantly monitoring the brightness of 100000 stars that are generally between 600 and 3000 light years from Earth. A regularly repeating dip in the brightness of a star indicates to scientists on the ground the possibility of a planet circling that star. They then look at the exact circumstances, such as how often the dip repeats, what fraction of the star's light is dimmed and for how long, to establish the size of the planet, its mass and how long it takes to circle its parent star.

The ultimate aim of the Kepler mission is to find planets of a similar size to Earth circling their parent stars at just the right distances so that they are not too hot or too cold for the possibility of life.

Scientists also use other techniques to find planets around other stars. The main one involves looking for tiny wobbles in the motion of stars due to planets moving around them. This technique has been spectacularly successful – thanks to it, we now know of nearly 500 planets outside our solar system – but, unlike the transit method, it has a bias towards massive planets circling close to the parent stars.

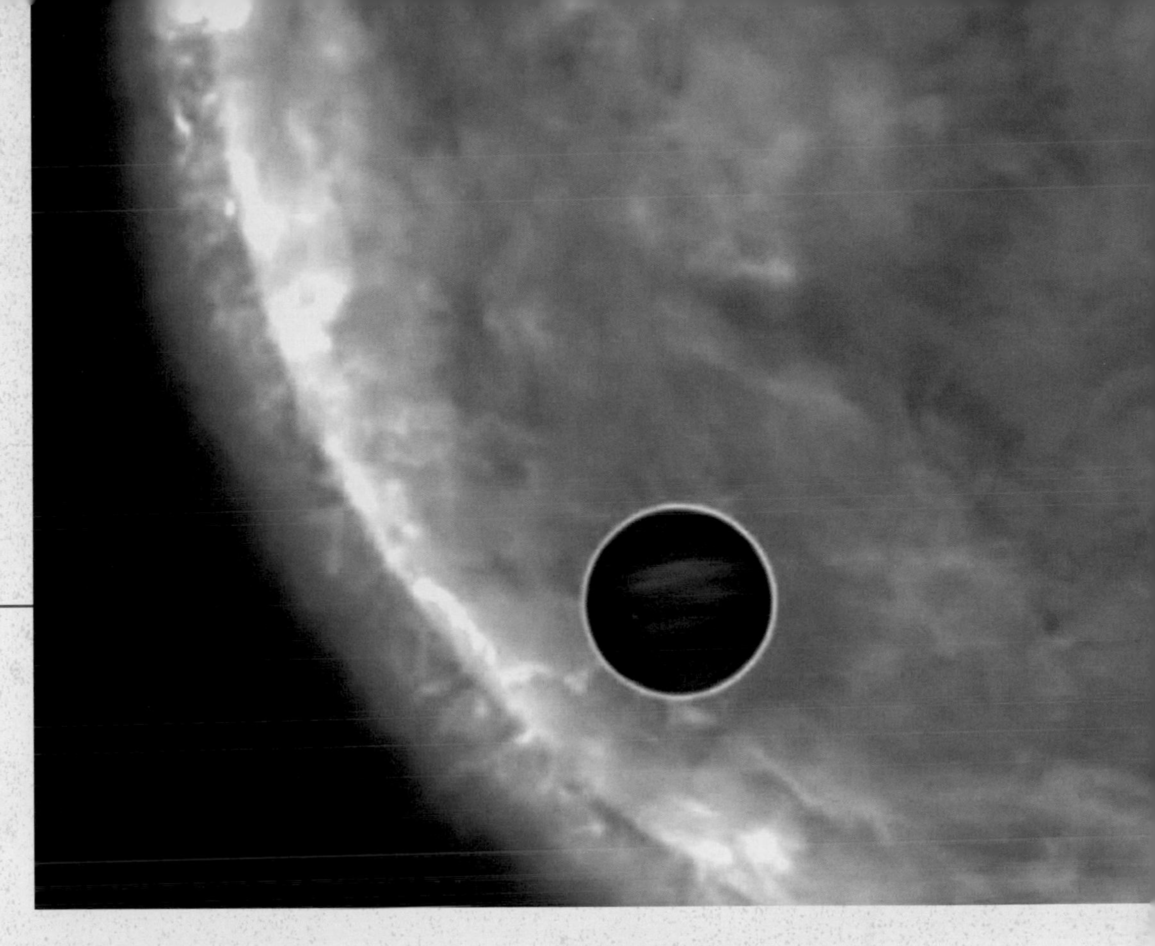

As of February 2011 Kepler had identified over 1200 possible planets. To qualify as definite planets, ground-based verification is needed, as well as follow-up observations from Kepler itself. Of the 1200 possible planets, 68 are about the same size as the Earth, and five of these Earth-sized planets circle their parent stars within the habitable zone, that is, the zone in which life as we know it could occur because liquid water could exist on the surface of the planet.

A recent discovery by Kepler is that of a yellow dwarf star, dubbed Kepler-11, with six planets circling it (an artist's impression of this solar system is shown overleaf). This system has little resemblance to our solar system – five of the planets circle their star at a distance closer than Mercury from the Sun, while the sixth planet is only a little further away. All of these planets are larger than the Earth, with the largest being comparable in size to the gas giants Uranus or Neptune in our solar system. The discovery of this system by the transit method adds to our knowledge of the possibilities for planets and planetary systems in the Universe. And it confirms the usefulness of the transit method as a search technique for planets around stars other than the Sun.

A transiting planet obscures just a tiny fraction of the light from its parent star, allowing astronomers to detect its presence. Artist's impression. NASA/ESA/G. Bacon.

Scientists using NASA's Kepler spacecraft have discovered a star, known as Kepler-11, with at least six planets circling around it. In this artist's impression three of the planets are crossing (transiting) in front of the star. NASA/Tim Pyle

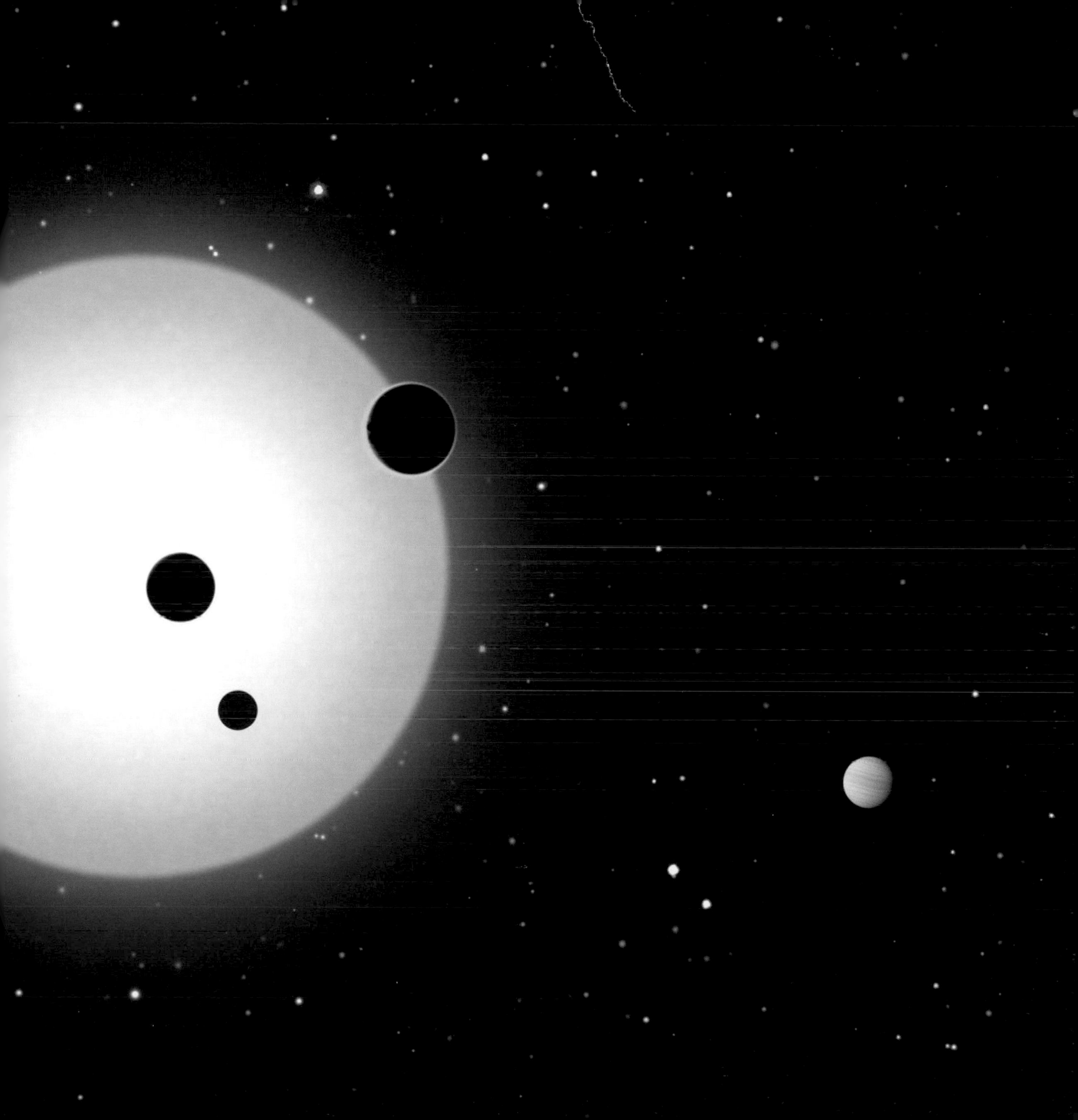

OBSERVING THE 2012 TRANSIT

2012

The last transit of Venus for the 21st century, and the last anybody now living is likely to see, will take place on 5 or 6 June 2012, depending on where it is being observed from. The entire transit will be visible from eastern Australia, New Zealand and parts of Asia. The transit will be visible in the United States during the afternoon for a few hours before sunset.

The transit of Venus photographed in 2004 by Xiaojin Zhu with an aeroplane flying across the face of the sun. Courtesy Xiaojin Zhu

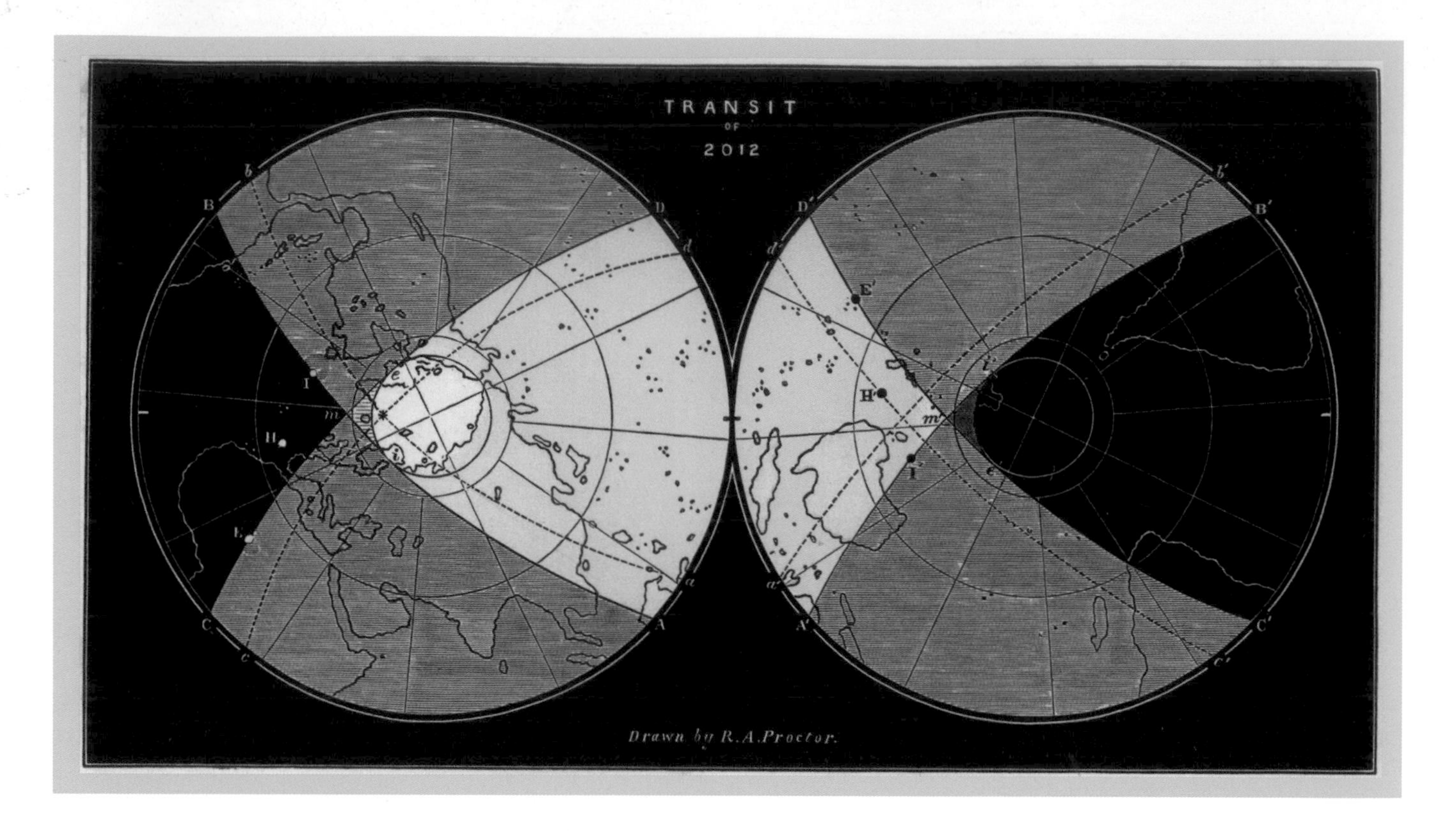

From the northern hemisphere the planet's motion during the transit is from left to right, that is, from east to west. What astronomers call 'first contact' is when Venus first touches the edge of the Sun, while 'second contact' is when Venus is fully on the disc of the Sun and just touching its edge. 'Third contact' is similar to the second, except that it takes place as Venus is about to move off the disc, and 'fourth contact' is when Venus last appears to touch the disc. During the transit of 2004 (when viewed from the northern hemisphere) Venus crossed the lower southern part of the Sun (as indicated in the photograph on the opposite page) while in 2012 it

∧ The visibility of the 2012 transit of Venus, with the northern hemisphere on the left and the southern hemisphere on the right. The brightest area denotes the regions from which the transit will be fully visible, the shaded areas indicate regions from which the beginning or end of the transit will be visible, and the darkest areas are the regions from which the transit will not be seen. Map from *Transits of Venus* by RA Proctor, 1874. Brian Greig collection

> A multi-exposure sequence of the 2004 transit taken through a 10-centimetre lens telescope in Wilden, Austria. Photo: Johannes Schedler, Panther Observatory

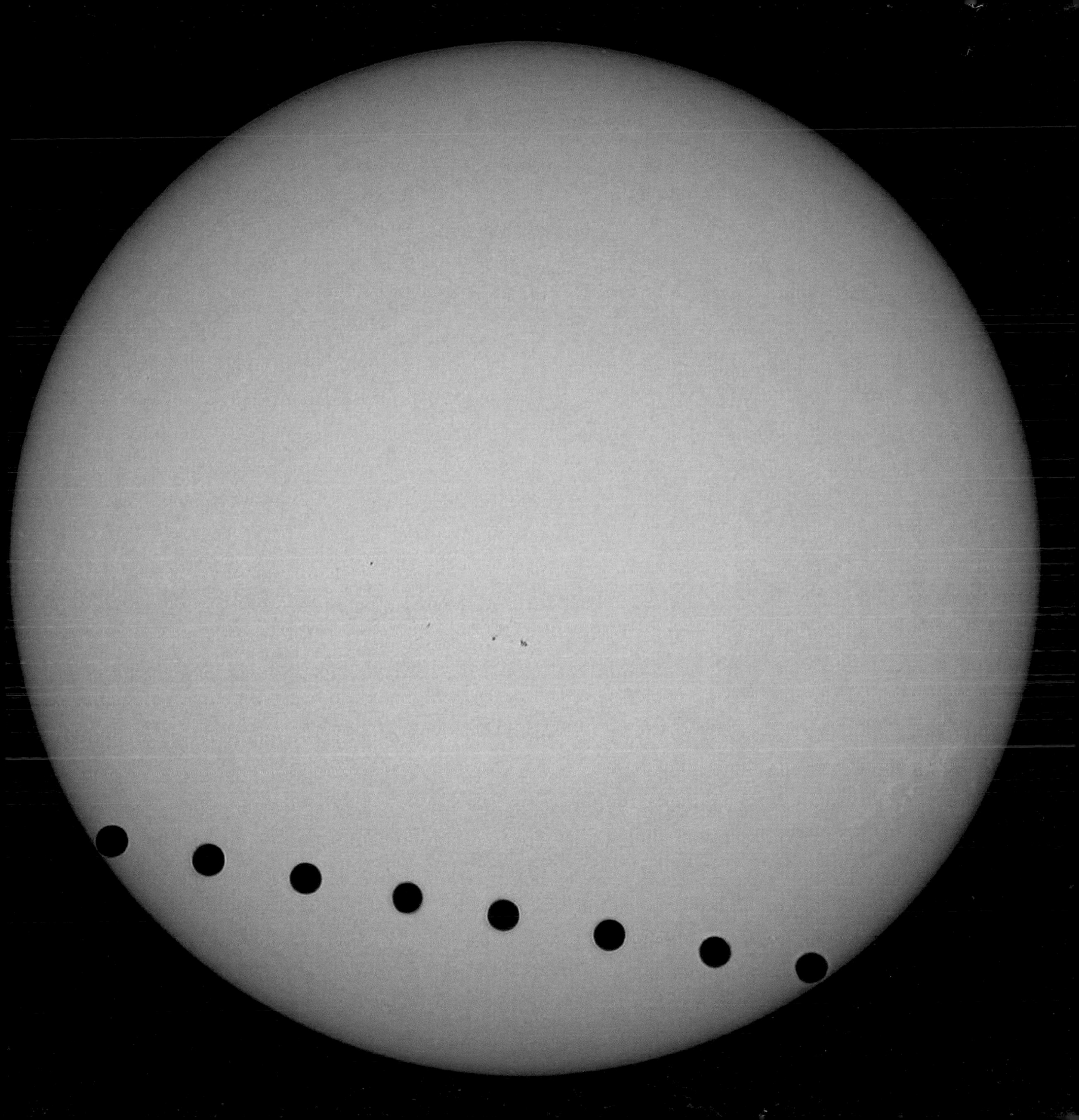

Southern hemisphere view

Northern hemisphere view

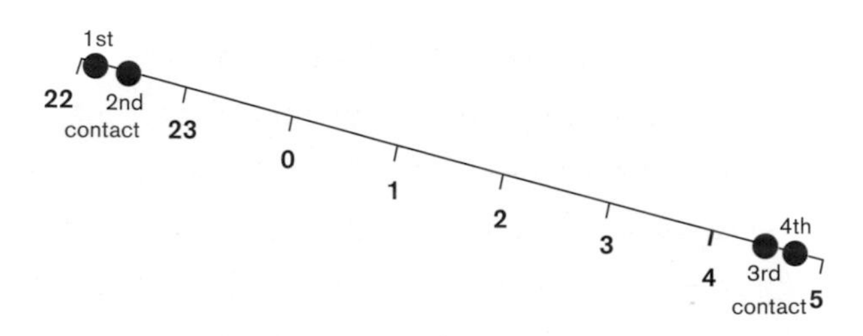

SUN

SUN

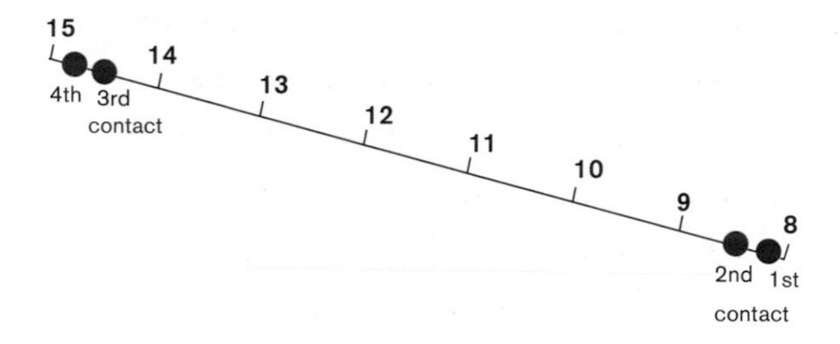

The path of Venus across the Sun as it will be seen from the southern hemisphere (left) and northern hemisphere (right) on 5–6 June 2012. First and fourth contacts refer to the times when Venus just touches the edge of the Sun's disc from the outside. Second and third contacts occur when Venus is just touching the edge from inside the disc. For the southern hemisphere the hourly marks are in Australian Eastern Standard Time (AEST), while for the northern hemisphere the hourly marks are in Universal Time (UT).

will cross the higher northern part. The diagrams on page 204 show the path of Venus across the Sun in both hemispheres.

For most of the United States the visibility of the transit in 2012 improves as one travels across the continent from east to west. Even from the east coast there will be several hours of visibility before sunset. In New York, for example, the transit will begin a little after 6 pm. As this is after work and after school, it is a convenient time for observation for many people. Weather permitting, there will be over two hours to see the transit before sunset. Of course, suitable locations with good sight-lines towards the west will need to be found.

Further west, from Chicago, the transit begins at just after 5 pm, giving over three hours of observation. On the west coast, from Los Angeles, the transit starts not long after 3 pm, providing almost five hours to see the event. The entire transit will be visible from Alaska and from Hawaii. From Honolulu, the Sun will be almost overhead at the start of the transit and near the horizon at the end. Those who would like to observe egress should locate a suitable observing site beforehand.

From the northern parts of South America the transit will still be in progress at sunset, while the rest of South America will not get a view (see the table on page 210 for detailed viewing information for North and Central American cities).

The best places to view the entire transit in other parts of the world will be from Japan, Korea, the eastern parts of China and the Russian Federation, Australia, New Zealand and New Guinea.

From Tokyo (Japan), Seoul (South Korea) and Hong Kong (China) the entire transit will be visible. In Tokyo and Seoul, the Sun will be well clear of the horizon for all contacts, but in Hong Kong the Sun will still be near the horizon at the start.

People on the west coast of Australia will miss the start of the transit, which will already be in progress there at sunrise. Although the transit will be visible from beginning to end in south and southeast Australia, such as Adelaide, Melbourne and Sydney, that does not mean that observing it will be easy. It will be winter in the southern hemisphere, so the Sun will rise late and set early. Consequently, at both the start and end of the transit the Sun may be low in the sky. This may make it necessary to find separate locations for viewing ingress and egress. Fortunately, there are almost six hours in between the two internal contacts to move between observing locations, if necessary.

From New Zealand the whole transit will also be visible except for some places on the east coast, where the Sun will set just before external egress.

Transits of Venus occur at regular, if infrequent, intervals. To understand this pattern we have to look at how the Earth and Venus circle the Sun.

SINCE THE FIRST USE OF THE telescope in the early 1600s, transits of Venus have occurred only a handful of times, in 1631 (unobserved), 1639 (the first observed), 1761, 1769, 1874, 1882 and 2004, and the next will be in 2012. It is clear that transits occur in pairs, eight years apart, separated by more than a century. Why this pattern? And why do transits of Venus always occur in early June or early December?

A transit can occur only if the Sun, Venus and the Earth are in the one line. This position is called conjunction or, more precisely, inferior conjunction when Venus and the Earth are on the same side of the Sun. Venus circles the Sun faster than the Earth, taking 224.701 days to be back in the same position in relation to the distant stars. The Earth, being farther from the Sun than Venus, takes a more leisurely 365.256 days, again with respect to distant stars, a period referred to as a sidereal year. From these periods, astronomers can calculate that, to reach one conjunction from the previous one, Venus will need to circle the Sun more than 2.5 times, while the Earth will go around the Sun about 1.6 times.

Now if Venus circled the Sun in the same plane as the Earth there would be a transit every 1.6 years, and it would be a common and mundane event. However, the plane of Venus' path is tilted to that of the Earth by just over 3 degrees. That means that at most inferior conjunctions Venus appears either above or below the Sun and is not crossing its disc.

A transit can only occur if Venus happens to be at a point in its path that crosses the plane of the Earth's orbit. The planes of the two orbits cross in a straight line – you can make an example of two planes meeting in a straight line by folding a piece of paper, as the bend is always a straight line. The straight line intersects the orbit of the Earth at two points, one point passed by the Earth in early June and the other in early December. Hence transits of Venus can only occur at those times of the year.

Let us assume that a transit takes place in early June. When is the next one? It can only happen after a whole number of years, so that the Earth is back in the original spot. We saw above that conjunctions happen every 1.6 years and so the lowest number of conjunctions to make a whole number of years is five, which makes up eight years. Hence eight years after the first conjunction there is another conjunction with the Earth back in its previous location at an intersection point between the two orbits. Another transit can then take place demonstrating that transits can occur at eight-year intervals.

Of course, the relationship that five conjunctions are equal to eight years is not exact. There is an error of about 2.5 days in the mathematical relationship, plus we have to take into account a slight shift in the points of intersection of the two orbital planes. Taken together, this means that if there is a conjunction just before Venus reaches the intersection point and there is a transit, eight years later there is likely to be another conjunction with Venus on the other side of the intersection point. Hence there can be two transits eight years apart, both at the same time of the year, say early June. Then there will be a long interval until the next time there is a conjunction near the next intersection point in December.

The pattern in the occurrence of transits is, starting from a June transit, eight years, 105.5 years, eight years, 121.5 years, and so on. This gives a total of 243 years for a complete cycle, as for example, from the June transit of 1769 to the June transit of 2012. There is a shorter time from a June transit to the following December transit than from a December transit to the next June transit. This is due to the slightly oval-shaped paths of the two planets so that their relative speeds vary at different parts of their orbits.

It needs to be pointed out that if a transit takes place with Venus exactly on the intersection point of the two orbits, then eight years later the conjunction occurs too far from that point for a transit to occur. Such a situation will not occur for a long time to come. When it does, the pattern of transits will change to the simpler one – 113.5 years, 129.5 years and so on, but the total 243 years for a full cycle is unchanged.

The paths of Venus and Earth around the Sun from side-on, showing the line of intersection.

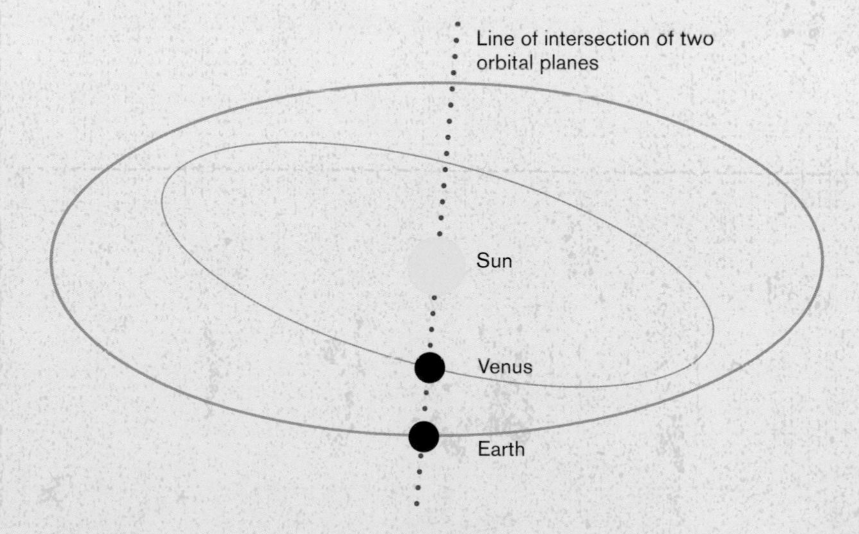

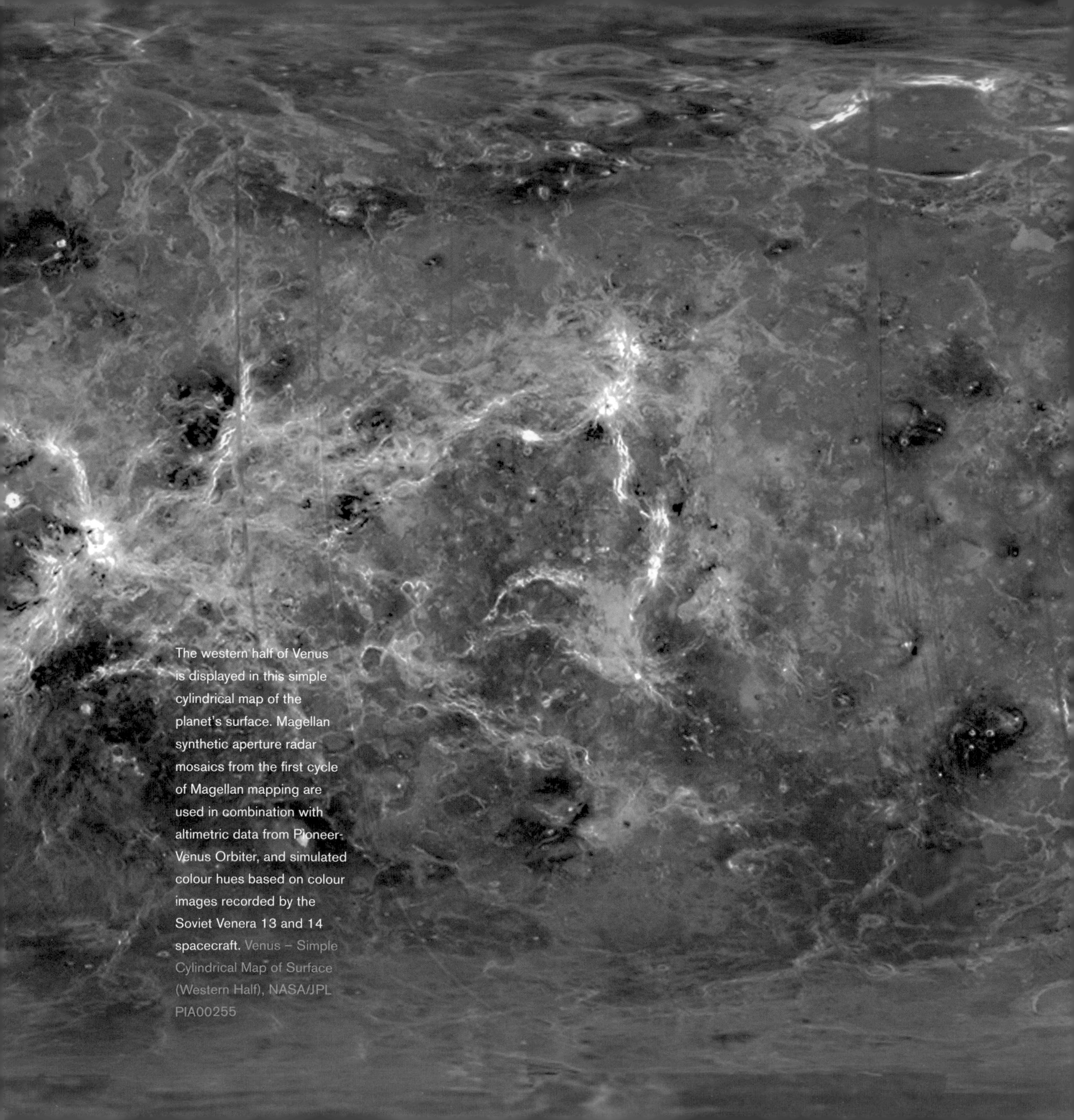

The western half of Venus is displayed in this simple cylindrical map of the planet's surface. Magellan synthetic aperture radar mosaics from the first cycle of Magellan mapping are used in combination with altimetric data from Pioneer-Venus Orbiter, and simulated colour hues based on colour images recorded by the Soviet Venera 13 and 14 spacecraft. Venus – Simple Cylindrical Map of Surface (Western Half), NASA/JPL PIA00255

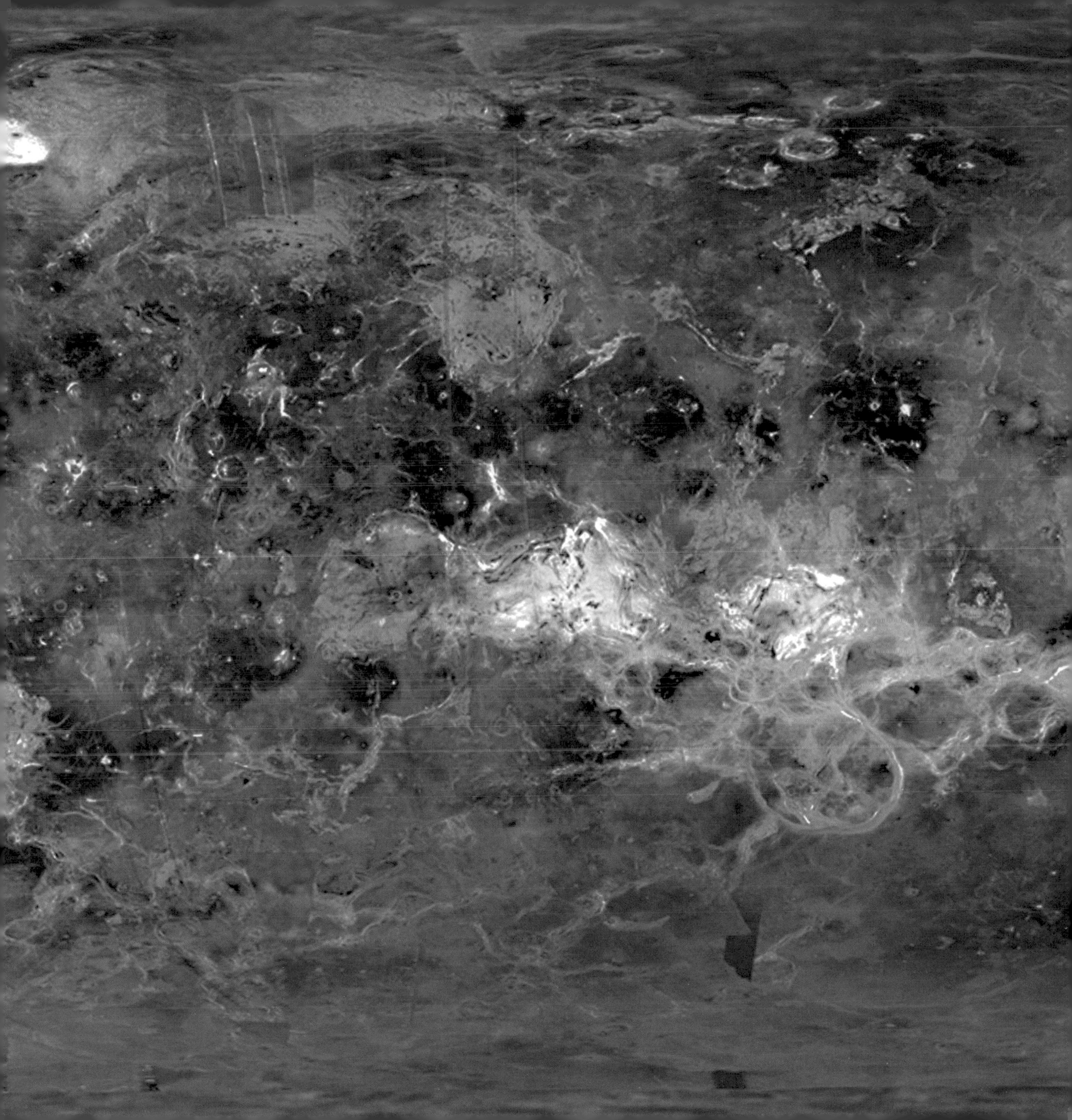

Obviously, if you just want to see Venus on the Sun during the transit, then any viewing location will do as long as it has a reasonable view of the sky. If you can find a location that is more likely to be free of clouds than other locations, that's even better.

Historically, the interesting parts of a transit of Venus are the part between first and second contact at the beginning (ingress) and the part between third and fourth contact at the end (egress). Those who want to try to see both the beginning and end of the transit will need to find a location from where the Sun can be seen at those phases of the transit. If such a location is not available, then two separate locations will be are needed for ingress and for egress.

For Europe (apart from parts of Spain and Portugal), the Middle East, eastern parts of Africa,

Viewing times for 2012 transit of Venus – North and Central America

Place	Time zone	1st contact	Elevation of Sun	2nd contact	3rd contact	4th contact	Elevation of Sun
Anchorage	AKDT	2:07 PM	51°	2:24 PM	08:31 PM	08:48 PM	14°
Baltimore	EDT	6:04 PM	26°	6:21 PM			
Chicago	CDT	5:04 PM	34°	5:22 PM			
Honolulu	HAST	12:10 PM	85°	12:28 PM	6:27 PM	6:44 PM	5°
Los Angeles	PDT	3:06 PM	58°	3:24 PM			
Miami	EDT	6:04 PM	26°	6:22 PM			
Mexico City	CDT	5:06 PM	41°	5:24 PM			
New York City	EDT	6:04 PM	24°	6:21 PM			
Salt Lake City	MDT	4:06 PM	52°	4:23 PM			
Toronto	EDT	6:04 PM	28°	6:21 PM			
Vancouver	PDT	3:06 PM	55°	3:23 PM			

This table shows the local time and elevation of the sun for contacts of the transit of Venus on 5 June 2012. See table on p. 214 or this website for other locations: <eclipse.gsfc.nasa.gov/transit/venus1412.html>.

This false-color image is a near-infrared map of lower-level clouds on the night side of Venus, obtained by the Near Infrared Mapping Spectrometer aboard the Galileo spacecraft as it approached the planet's night side. The spacecraft is about 100000 kilometres above the planet. The red colour represents the radiant heat from the lower atmosphere shining through the sulphuric acid clouds, which appear as much as ten times darker than the bright gaps between clouds. Infrared Image of Low Clouds on Venus, NASA/JPL PIA00124

How to safely observe the last transit of Venus until 2117.

The safest way to watch the transit is to project the image through a telescope onto a white screen. Illustration: Nick Lomb & Peter Thorn Design

It has to be emphasised that it is dangerous to look directly at the Sun at any time, even when it is low in the sky. Serious and permanent eye damage can result from allowing direct sunlight into the eye. As the eye does not have pain receptors this damage only becomes evident after the event. Specialist astronomy and telescope shops sell filters that can be placed at the front of telescopes and cardboard 'eclipse shades' for the unaided eye. These should be used with great care and children need to be supervised at all times. Homemade filters are unacceptable as they could allow damaging ultraviolet and/or infrared radiation into the eye.

If you have a small telescope or a pair of binoculars, the safest way to observe the transit is by projection. With your back to the Sun, project the image of the Sun on a piece of white card. During the transit, Venus will be clearly seen as a black dot slowly moving across the face of the Sun. The black dot will appear clearly as the apparent size of Venus at the time will be a reasonably large fraction, one-thirty-third, of the size of the Sun.

Do not give up if you do not have your own telescope or binoculars because there will be plenty of public observing sites. Contact your local public observatory, planetarium or amateur astronomy group for details. Where the transit begins early in the morning, as in Australia, you may be able to view at a public session before work or school. Alternatively, you may consider that it is worth taking some time off to see the last transit of Venus until 2117.

The 2004 transit of Venus photographed through clouds from a lookout above the Catawba River near Connelly's Springs, North Carolina, USA.
Photo: David Cortner

India and Indonesia, the transit will be in progress at sunrise. So to view the transit from its beginning in, say, London it will be necessary to rise early and find a suitable spot from which the Sun can be seen when it is just 8° above the eastern horizon. The western parts of Africa will not get a view of the transit.

For all other places not covered in the table below, see NASA's 2004 and 2012 transits of Venus website <eclipse.gsfc.nasa.gov/transit/venus0412.html>.

If at least part of the 2012 transit of Venus is visible from your location, make an effort to view it and do so safely. This is your last chance to see one!

Viewing times for 2012 transit of Venus – Rest of the World

Place	Time zone	Day	June	1st contact	Elevation of Sun	2nd contact	3rd contact	4th contact	Elevation of Sun
ASIA									
Hong Kong	HKT	Wednesday	6	6.12 am	6°	6.30 am	12.31 pm	12.49 pm	84°
Seoul	KST	Wednesday	6	7.10 am	21°	7.28 am	1.31 pm	1.48 pm	68°
Singapore	SGT	Wednesday	6				12.31 pm	12.49 pm	68°
Tokyo	JST	Wednesday	6	7.11 am	31°	7.28 am	1.30 pm	1.47 pm	59°
EUROPE									
Helsinki	EEST	Wednesday	6				7.37 am	7.54 am	24°
London	BST	Wednesday	6				5.37 am	5.55 am	8°
Moscow	MSD	Wednesday	6				8.37 am	8.54 am	31°
Rome	CEST	Wednesday	6				6.38 am	6.55 am	13°
OCEANIA									
Auckland	NZST	Wednesday	6	10.16 am	24°	10.33 am	4.25 pm	4.43 pm	4°
Papeete	TAHT	Tuesday	5	12.13 pm	50°	12.30 pm			
Sydney	AEST	Wednesday	6	8.16 am	13°	8.34 am	2.26 pm	2.44 pm	20°

This table indicates the local time and elevation of the Sun for contacts of the transit of Venus on 5 or 6 June 2012. For all places not covered in the table see this website: <eclipse.gsfc.nasa.gov/transit/venus1412.html>.

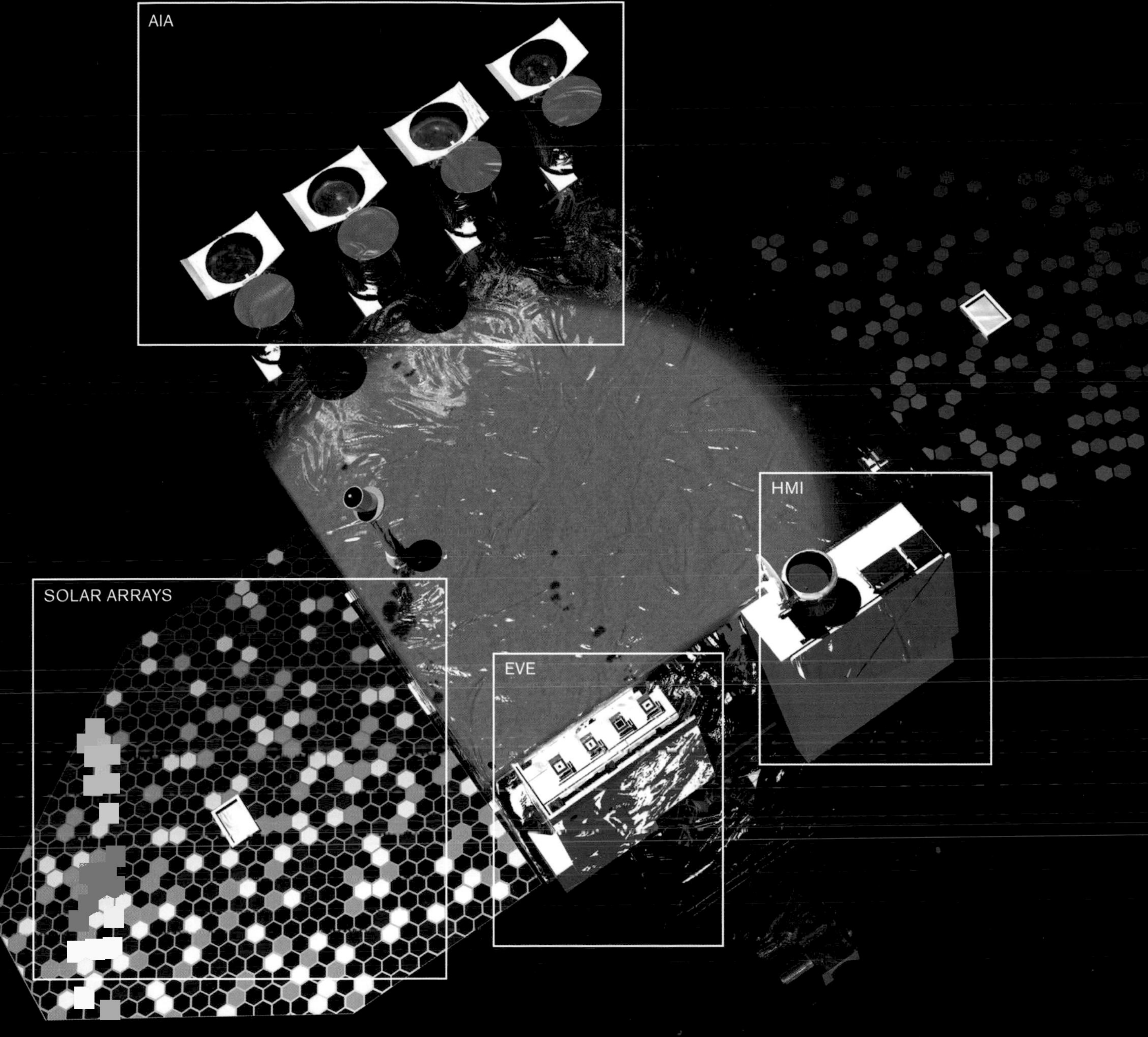

In 2012 NASA's Solar Dynamics Observatory spacecraft (SDO) will replace TRACE in observing the 2012 transit. SDO has three main instruments: Atmospheric Imaging Assembly (AIA) to image the Sun in different colours; Extreme Ultraviolet Variability Experiment (EVE) to measure extreme ultraviolet radiation; and Helioseismic and Magnetic Imager (HMI) to map vibration and magnetic fields across the Sun's disc. The solar arrays power the spacecraft's systems, while the high-gain antennas keep the probe in contact with Earth. NASA

GLOSSARY

Altimeter (radar) Carried on board a satellite, an instrument that uses radar waves to measure the heights of surface features below.

Aperture (telescope) The width of the mirror or main lens on a telescope. This is the key dimension for a telescope as the sharpness of the image depends directly on the aperture, while the quantity of light the telescope can collect depends on the square of the aperture.

Bark - barque (ship) A ship with a flat and broad front or bow and a square stern.

Black drop An optical effect that disturbed many 18th- and 19th-century astronomers trying to time the instants that Venus touched the edge of the Sun during a transit. A common manifestation was as a dark thread joining the edges of the planet and the Sun.

Chronometer A highly accurate and portable clock designed to work despite the motion and temperature changes experienced on board a ship.

Clock (error rate) The time lost or gained each day by a clock.

Clock (pendulum) A clock in which the time keeping is controlled by the side-to-side motion of a pendulum.

Comet A body composed of ice, rock and dust that orbits the Sun on an elongated path. Some of the ice (water, carbon dioxide and other frozen gases) on its surface vaporises and releases gas and dust as it approaches the vicinity of the Sun.

Conjunction (inferior) This occurs when either the planet Mercury or the planet Venus lies between the Earth and the Sun.

Contacts There are four contacts during a transit: first contact when the planet touches the edge of the Sun from the outside before starting to move in front of it, second contact when it touches the edge from the inside, third contact when it touches the edge of the Sun just before moving off the Sun and fourth contact when it last touches the edge.

Dead reckoning A method of deducing the position of a vessel from its last known position and its estimated speed and direction.

Eclipse An event in which one astronomical object is covered by another moving in front of it, such as in an eclipse of the Sun when it is covered by the Moon. Alternatively, an event in which an object moves into the shadow of another, such as in an eclipse of the Moon when it moves into the Earth's shadow.

Egress The last stages of a transit when a transiting body (such as Venus) is leaving the body (such as the Sun) in front of which it has been moving.

Filter (eyepiece) A dark piece of glass placed in front of the eyepiece of a telescope to reduce the heat and light intensity from the Sun. Such filters are now regarded as highly dangerous as they may crack due to the heat and are banned in Australia.

Filter (full aperture) A filter placed at the far end of the telescope tube preventing most of the Sun's heat and light from entering the telescope and reaching the eye or camera.

Filter (hydrogen alpha) A highly specialised filter for a telescope that blocks all light except for a very narrow range of red light emitted by hydrogen atoms.

Gravity The weakest of the four known forces of Nature, yet the one that is most important over large distances. It holds us to the surface of the Earth and keeps the Earth in its circular path around the Sun.

Heliostat A mirror driven by clockwork that can track the daily motion of the Sun and reflect its light in a specified direction.

Ingress The first stages of a transit when a transiting body (such as Venus) is moving in front of the body (such as the Sun) that it will cross.

Janssen apparatus A device based on a design by French astronomer Jules Janssen that can take a series of images one after the other on a circular photographic plate.

Latitude A coordinate indicating position in degrees north or south of the equator and one of the two coordinates needed to specify position on the Earth's surface.

Limb darkening The slightly darker appearance of the disc of the Sun near its edge. The effect is due to the Sun being a sphere – near its edge we are looking at higher, cooler and hence darker gases.

Longitude Coordinate indicating position in degrees east or west of a predetermined place, usually taken to be Greenwich Observatory near London, and one of the two coordinates needed to specify position on the Earth's surface.

Lunar distances A method to establish time at Greenwich, and hence longitude, by measuring the angular distances of bright stars from the Moon.

Mappe monde Two hemispheric maps together covering the whole globe.

Meridian An imaginary line in the sky that passes north–south and overhead.

Meteoroid A particle or small body that has been circling the Sun and hits the Earth's atmosphere. As the particle burns up high in the atmosphere a *meteor* can be seen. If part of the particle or body survives and hits the ground then it is a *meteorite*.

Micrometer A device to measure small angular separations (say between two stars) in the field of view of a telescope eyepiece.

Nebula A blurry or fuzzy object in the sky that could be a relatively nearby cloud of gas and dust or it could be a distant galaxy.

Parallax The angle formed or subtended by the radius of the Earth as it would be seen from a distant object. For example, the Sun and the Moon subtend an angle of half a degree as seen from Earth, while an outstretched hand at arm's length subtends about 22° at the eye. A simple trigonometrical formula relates the parallax of an object to its distance from the Earth. Astronomers observing transits in past centuries were trying to measure the parallax of the Sun as a step towards obtaining its distance.

Photoheliograph A specialised 19th-century telescope designed to take photographs of the Sun.

Planet A large body that circles the Sun in its own unique path around the Sun and shines by light reflected from the Sun.

Planet (dwarf) A body circling the Sun that is large enough for gravity to pull it into a circular shape, but not large enough to be called a planet.

Planet (minor) A body circling the Sun that is not large enough to have enough gravity to assume a circular shape.

Quadrant An instrument with which astronomers could measure the angular elevation of a star or the Sun from the horizon.

Radar A technique to map a surface by bouncing radio waves off it.

Seeing A technical term in astronomy providing a measure of the amount of blurring of a star image due to the passage of its light through the Earth's turbulent atmosphere.

Sextant An instrument for use on board a ship to measure the angular elevation of stars or the Sun from the horizon.

Spectrum The range of colours observed when a prism or another device breaks light into components. A rainbow is an example of a spectrum; it is due to raindrops spreading sunlight into its component colours.

Spectrum (secondary) Lens telescopes normally use an achromatic main lens that bring two colours, say, red and blue, to the one focus. Small blurring can still occur from the remaining colours not quite coming to the same focus. This blurring is called the secondary spectrum.

Telegraph A method of communicating over large distances by sending electrical signals over wires. The telegraph was the 'internet' of the late 19th and early 20th centuries. Messages were encoded into dots and dashes by the sender and decoded by the receiver.

Telescope (achromatic) A lens telescope with an objective lens made of two or more pieces of different types of glass that allow the lens to bring light of different colours to the same focus.

Telescope (finder) A smaller telescope with a wide field of view mounted on a larger one so as to make it easier to locate objects to view.

Telescope (lens) A telescope that collects light using a large lens or objective. It is also known as a refracting telescope.

Telescope (mirror) A telescope that collects light using a curved mirror that reflects and focuses light towards the eyepiece. It is also known as a reflecting telescope.

Telescope (transit) A telescope constrained to move in the north–south direction only and designed to allow astronomers to time when stars or the Sun cross the overhead meridian.

Transit This occurs when a body such as Venus passes in front of, or transits, the disc of a star such as the Sun.

Year (sidereal) The time taken by a planet such as the Earth to circle the Sun and return to its original position with respect to distant stars.

BIBLIOGRAPHY

INTRODUCTION

Clerke, AM (1902) *A Popular History of Astronomy During the Nineteenth Century*, fourth edition, A&C Black, London.

Harkness, W (1883) 'Address', in *Proceedings of the American Association for the Advancement of Science, Thirty-First Meeting, Held at Montreal, Canada*, August, 1882, 77.

Hughes, DW (2001) 'Six stages in the history of the astronomical unit', *Journal of Astronomy History and Heritage*, 4: 15–28.

Jodrell Bank Observatory (2004) *Measurement of the Distance of Venus by Radar*, viewed 21 February 2011, <jb.man.ac.uk/~slowe/transit2004/science_dist_radar.html>.

Lomb, N (2004) *Transit of Venus: The Scientific Event that Led Captain Cook to Australia*, Powerhouse Publishing, Sydney.

NASA Goddard Space Flight Center (2010) *Venus Fact Sheet*, viewed 21 February 2011, <nssdc.gsfc.nasa.gov/planetary/factsheet/venusfact.html>.

Pasachoff, JM, Schneider, G & Widemann, T (2011) 'High-resolution satellite imaging of the 2004 transit of Venus and asymmetries in the Cytherean atmosphere', *Astronomical Journal*, in press, viewed 21 February 2011, <web.williams.edu/Astronomy/eclipse/transits/AJ_ToV_2010_&data.pdf>.

US Nautical Almanac Office (2010) *Astronomical Almanac 2011*, viewed 21 February 2011, <asa.usno.navy.mil/SecK/2011/Astronomical_Constants_2011.pdf>.

A SPOT OF UNUSUAL MAGNITUDE: 1639

Brewster, D (1841) *The Martyrs of Science, or the Lives of Galileo, Tycho Brahe, and Kepler*, John Murray, London.

Chapman, A (1990) 'Jeremiah Horrocks, the transit of Venus, and the "New Astronomy" in early seventeenth-century England', *Quarterly Journal of the Royal Astronomical Society*, 31: 333–57.

Field, JV 'Johannes Kepler', in *The MacTutor History of Mathematics Archive*, viewed on 21 February 2011, <www-history.mcs.st-and.ac.uk/Biographies/Kepler.html>.

Proctor, RA (1874) *Transits of Venus. A Popular Account of Past and Coming Transits from the First Observed by Horrocks A.D. 1639 to the Transit of 2012*. Longmans, Green, and Co., London.

Van Helden, A (1976) 'The importance of the transit of Mercury of 1631', *Journal for the History of Astronomy*, 7:1–10.

Whatton, AB (1859) *The Transit of Venus Across the Sun: A Translation of the Celebrated Discourse Thereupon by the Rev. Jeremiah Horrox*, William MacIntosh, London, viewed 27 February 2011, <dlib.stanford.edu:6521 /text1/dd-ill/transit-memoir.pdf>.

FROZEN PLAINS AND TROPICAL SEAS: 1761

Chaix d'Est-Ange, GL (1910) *Dictionnaire des Familles Françaises Anciennes ou Notables à la Fin du XIXe Siècle. IX. Cas-Cha*, Charles Hérrissey, Évreux, 378–79.

Chappe d'Auteroche, J (1770) *A Journey into Siberia Made by the Order of the King of France*, translated from the French, T. Jefferys, London.

Fernie, JD (1997) 'Transits, travels and tribulations, II', *American Scientist*, 85: 418–21

—— (1998) 'Transits, travels and tribulations, II', *American Scientist*, 86: 123–126

Halley, E (1716) 'Containing Doctor Halley's Dissertation on the method of finding the Sun's parallax and distance from the Earth, by the transit of Venus over the Sun's Disc, June the 6th, 1761. Translated from the Latin in Motte's Abridgement of the Philosophical Transactions, Vol, I. pag. 243; with additional notes.' In James Ferguson, *Astronomy Explained upon Sir Isaac Newton's Principles*, second American edition, Philadelphia 1806, 480–95.

Institut de France Académie des sciences, 'Liste des membres, correspondants et associés étrangers de l'Académie des sciences depuis sa création en 1666', viewed 23 February 2011, <academie-sciences.fr/academie/membre/memC.pdf>.

Le Gentil de la Galaisière, GJHJB (1779)

Voyage dans les Mers de l'Inde: Fait par Ordre du Roi, à l'occasion du Passage de Vénus sur le Disque du Soleil le 6 Juin 1761, et le 3 du même mois 1769, Royale, Paris.

MacKenzie, T (1951) 'Mason and Dixon at the Cape', *Monthly Notes of the Astronomical Society of South Africa* 10: 99–102.

Hogg, HS (1951) 'Le Gentil and the transits of Venus, 1761 and 1769', *Journal of the Royal Astronomical Society of Canada*, 45: 37–44.

Proctor, RA (1874) *Transits of Venus. A Popular Account of Past and Coming Transits from the First Observed by Horrocks A.D. 1639 to the Transit of 2012*. Longmans, Green, And Co., London.

Short, J (1761) 'An account of the transit of Venus over the Sun, on Saturday morning, 6th June 1761, at Savile-House, about 8" of time west of St Paul's, London', *Philosophical Trans-actions of the Royal Society* LII: 178–82.

—— (1762) 'The observations of the internal contact of Venus with the Sun's limb, in the late Transit, made in different places of Europe, compared with the time of the same contact observed at the Cape of Good Hope, and the parallax of the Sun, from thence determined', *Philosophical Transactions of the Royal Society* LII: 613–28.

VENUS OF THE SOUTH SEAS: 1769

Beaglehole, JC (1974) *The Life of Captain James Cook*, Adam & Charles Black, London.

Bray, RJ (1980) 'Australia and the transit of Venus', *Proceedings Astronomical Society of Australia*, 4: 114–20.

Chappe d'Auteroche, J (1778) *A Voyage to California: To Observe the Transit of Venus*, translated from the French, Edward & Charles Dilly, London.

Cook, J (1769) *Journal of the H.M.S. Endeavour, 1768–1771*, National Library of Australia, MS 1, transcript, viewed on 24 February 2011, <southseas.nla.gov.au/journals/cook/contents.html>.

Fernie, JD (1998) 'Transits, travels and tribulations, IV', *American Scientist*, 86: 422–25.

—— (1999) 'Transits, travels and tribulations, V', *American Scientist*, 87: 119–21.

Green, C & Cook, J (1771) 'Astronomical Observations made, by Appointment of the Royal Society, at King George's Island in the South Sea', *Philosophical Transactions of the Royal Society*, 61: 397–421.

Morris, M (1980) 'Man without a face – Charles Green', viewed on 24 February 2011, <captaincooksociety.com/ccsu4143.htm>.

Schneider, G, Pasachoff, JM & Golub, L (2001) 'TRACE observations of the 15 November 1999 transit of Mercury', presented at the 33rd meeting of the AAS Division for Planetary Sciences, New Orleans, Louisiana, viewed on 23 February 2011, <nicmosis.as.arizona.edu:8000/POSTERS/TOM1999.jpg>.

CAPTURING THE TRANSIT: 1874 AND 1882

Anonymous (1874a) 'Wednesday, December 9, 1874', *Argus*, 9 December, 4.

—— (1874b) 'Our Melbourne letter', *Queenslander*, 26 December, 2.

—— (1874c) 'Death of Professor Wilson', *Argus*, 30 December, 2.

—— (1882) 'Venus on the Sun's disk', The Sun, 7 December, 1.

—— (1903) 'Astronomer Harkness dead', *New York Times*, 1 March.

—— (2007) '1882 transit of Venus', HM Nautical Office, viewed 24 March 2011, <www.hmnao.com/nao/transit/V_1882/>

Ball, RS (1886) *The Story of the Heavens*, Cassell & Co., London, 210.

Chambers, GF (1889) *A Handbook of Descriptive and Practical Astronomy: I The Sun, Planets, and Comets*, fourth edition, Clarendon Press, Oxford, 457–67.

Clark, BAJ & Orchiston, W (2004) 'The Melbourne Observatory Dallmeyer photoheliograph and the 1874 transit of Venus', *Journal of Astronomy History and Heritage*, 7: 44–49.

Council of the Royal Astronomical Society, (1884) 'The transit of Venus, 1882', *Monthly Notices of the Royal Astronomical Society*, 44: 182–84.

Crommelin, ACD (1923) 'George Lyon Tupman obituary', *Monthly Notices of the Royal Astronomical Society*, 83: 247–48.

Davidson, G (1881) 'Observations made at Nagasaki'. Section II in Simon Newcomb (ed) *Observations of the Transit of Venus, December 8–9, 1874, Made and Reduced under the Direction of the Commission Created by Congress, Part II*, page proof in US Naval Observatory Library

Débarbat, S & Launay, F (2006) 'The 1874 transit of Venus observed in Japan by the French, and associated relics', *Journal of Astronomical History and Heritage*, 9: 167–71.

De La Rue, W (1874) 'On a piece of apparatus for carrying out M. Janssen's method of time–photographic observations of the transit of Venus', *Monthly Notices of the Royal Astronomical Society*, 34: 347–53.

Dick, SJ (2005) 'The American transit of Venus expeditions of 1874 and 1882'. In DW Kurtz (ed) *New Views of the Solar System and Galaxy, Proceedings IAU Colloquium No. 196, 2004*, Cambridge University Press, Cambridge, 100–10.

Edwards, PG (2004) 'Charles Todd's observations of the transits of Venus', *Journal of Astronomy History and Heritage*, 7: 1–7.

Ellery, RLJ (1875) 'From a letter of Mr. Ellery, dated Melbourne Observatory, Dec. 30, 1874', *Monthly Notices of the Royal Astronomical Society*, 35: 216–17.

—— (1876) 'Notes on some of the physical appearances observed in the late transit of Venus', *Transactions and Proceedings of the Royal Society of Victoria*, 12: 60–65.

—— (1882) 'Observations of the transit of Venus, 1874, December 8–9, Colony of Victoria, Australia', *Reports of Observations of the Transit of Venus, 1874, December 8–9, Made in Victoria, New South Wales, South Australia, India, and the Cape of Good Hope*, Royal Astronomical Society, London, 31–47.

Ellyard, D (2005) *Who Discovered What When*, Reed New Holland.

Gascoigne, SCB (1992) 'Robert L.J. Ellery, his life and times', *Proceedings of the Astronomical Society of Australia*, 10: 170–76.

Harkness, W (1881) 'Observations made at Hobart town'. Section V in Simon Newcomb (ed) *Observations of the Transit of Venus, December 8–9, 1874, Made and Reduced under the Direction of the Commission Created by Congress, Part II*, page proof in US Naval Observatory Library.

—— (1891) 'The solar parallax and its relational constants', in *Observations made during the year 1885 at the U.S. Naval Observatory, Appendix III*, Government Printing Office, Washington, 51–55.

Hingley, PD (2005) 'The priest and the stuffed penguin', *Journal of the British Astronomical Association*, 115: 150–70.

Janiczek, PM (1974) 'Transits of Venus and the American expedition of 1874', *Sky & Telescope*, December, 366.

Launay, F & Hingley, PD (2005) 'Jules Janssen's "Revolver photographique" and its British derivative, "The Janssen slide"', *Journal for the History of Astronomy*, 36: 57–59.

Orchiston, W & Buchenan, A (1993) 'Illuminating incidents in antipodean astronomy: Campbell Town, and the 1874 transit of Venus', *Australian Journal of Astronomy*, 5: 11–31.

Orchiston, W & Buchenan, A (2004) '"The Grange", Tasmania: survival of a unique suite of 1874 transit of Venus relics', *Journal of Astronomical History and Heritage*, 7: 34–43.

Orchiston, W, Love, T & Dick, SJ (2000) 'Refining the astronomical unit: Queenstown and the 1874 transit of Venus', *Journal of Astronomical History and Heritage*, 3: 23–44.

Peters, CHF (1881) 'Observations made at Queenstown'. Section VII in Simon Newcomb (ed) *Observations of the Transit of Venus, December 8–9, 1874, Made and Reduced under the Direction of the Commission Created by Congress, Part II*, page proof in US Naval Observatory Library.

Raymond, CW (1871) 'Report of a reconnaissance of the Yukon River, Alaska Territory, July to September, 1869', Government Printing Office Washington.

—— (1881) 'Observations made at Campbell Town'. Section VI in Simon Newcomb (ed) *Observations of the Transit of Venus, December 8–9, 1874, Made and Reduced under the Direction of the Commission Created by Congress, Part II*, page proof in US Naval Observatory Library.

Ratcliff, J (2008) *The Transit of Venus Enterprise in Victorian Britain*, Pickering & Chatto, London.

Russell, HC (1892) *Observations of the Transit of Venus, 9 December, 1874; Made at Stations in New South Wales*, Government Printer, Sydney.

Symes, GW (1976) 'Todd, Sir Charles (1826–1910)', *Australian Dictionary of Biography Online Edition*, viewed 25 February 2011, <adbonline.anu.edu.au/biogs/A060301b.htm>.

Tebbutt, J (1861) 'A comet visible', *Sydney Morning Herald*, 25 May, 5.

—— (1874) 'Observations of the transit of Venus, 1874, December 8–9, made at Windsor, N.S. Wales, *Reports of Observations of the Transit of Venus,*

1874, December 8–9, Made in Victoria, New South Wales, South Australia, India, and the Cape of Good Hope, Royal Astronomical Society, London, 90–92.

—— (1883) 'Note on Professor Newcomb's remarks on the Windsor observations of the transit of Venus in 1874', *Monthly Notices of the Royal Astronomical Society*, 43: 279–80.

—— (1986) *Astronomical Memoirs*, Hawkesbury Shire Council, Windsor.

Todd, C (1882) 'Observations of the transit of Venus, 1874, December 8–9, at Adelaide, South Australia', *Reports of Observations of the Transit of Venus, 1874, December 8–9, Made in Victoria, New South Wales, South Australia, India, and the Cape of Good Hope*, Royal Astronomical Society, London, 93–96.

Todd, DP (1881) 'The solar parallax as derived from the American photographs of the transit of Venus, 1874, December 8–9', *The Observatory*, 4: 202–05.

Tupman, GL (1878) 'On the mean solar parallax as derived from the observations of the transit of Venus, 1874', *Monthly Notices of the Royal Astronomical Society*, 38: 429–57.

Tupman, GL (1881) 'Expedition to the Hawaiian (Sandwich) Islands', Part I in Sir George Airy (ed) *Account of Observations of the Transit of Venus, 1874, December 8, Made under the Auspices of the British Government and the Reduction of the Observations.*

Turner, HH (1890) 'Father Perry', *The Observatory*, 13: 81–86.

Watson, JC (1881) 'Observations made at Peking', Section III in Simon Newcomb (ed) *Observations of the Transit of Venus, December 8–9, 1874, Made and Reduced under the Direction of the Commission Created by Congress, Part II*, page proof in US Naval Observatory Library.

White, EJ (1878) 'Account of the telegraphic determination of the difference in longitude between Melbourne and Hobart town in the year 1875', *Transactions and Proceedings of the Royal Society of Victoria*, 13: 94–98.

SPACE-AGE TRANSIT: 2004

Dayton, L (2004) 'Rare transit of Venus trips a light speck-tacular', *Australian*, 9 June, 3.

James, N (2005) 'Sharm el Sheikh, Egypt', *Journal of the British Astronomical Association*, 115: 159–60.

Lomb, N & George, M (2003) 'Transit of Venus 8 June 2004', ASA Factsheet No. 15, Transit of Venus 8 June 2004, viewed 25 February 2011, <www.astronomy.org.au/ngn/media/client/factsheet_15.pdf>.

Macdonald, P (2002) 'The transit of Venus on 2004 June 8', *Journal of the British Astronomical Association*, 112: 319–24.

Macey, R (2004) 'Hundreds queue for a Captain Cook', *Sydney Morning Herald*, 9 June, 10.

Mewhinney, M, Hoover, R & Perrotto, TJ (2011) 'NASA's Kepler spacecraft discovers extraordinary new planetary system', viewed on 25 February, <nasa.gov/mission_pages/kepler/news/new_planetary_system.html>.

Mobberley, M (2005) 'Selsey, Sussex', *Journal of the British Astronomical Association*, 115: 140.

Pasachoff, JM, Schneider, G & Widemann, T (2011) 'High resolution satellite imaging of the 2004 transit of Venus and asymmetries in the Cytherean atmosphere', *Astronomical Journal*, 141, 112–120, viewed 21 February 2011, <web.williams.edu/Astronomy/eclipse/transits/AJ_ToV_2010_&data.pdf>.

Pazmino, J (2004) 'Venus on the Sun', Observer reports archive, viewed 28 April 2011, <aaa.org/observerreportsarchive>.

Scharmer, GB, Bjelksö, K, Korhonen, T, Lindberg, B & Petterson, B (2003) 'The 1-meter Swedish solar telescope'. In Stephen L. Keil and Sergey V. Avakyan (eds) *Innovative Telescopes and Instrumentation for Solar Astrophysics*, Proceedings of the SPIE, 4853: 341–50.

OBSERVING THE 2012 TRANSIT

Espanak, F (2002) '2004 and 2012 transits of Venus', *NASA Eclipse Website*, viewed 25 February 2011, <eclipse.gsfc.nasa.gov/transit/venus0412.html>.

Lomb, N & George, M (2011) 'Transit of Venus 6 June 2012', ASA Factsheet No. 24 Transit of Venus 6 June 2012, <astronomy.org.au/ngn/engine.php?SID=1000011>.

Venus viewed through violet (upper image) and near infrared filters (lower image).
NASA PIA00073

ACKNOWLEDGMENTS

I would like to thank Brian Greig, the orrery maker, who freely made available numerous illustrations that he had carefully collected over several decades, as well as providing information on the 1874 transit of Venus expeditions. Thanks are also due to Martin George, who literally went out of his way to take new images of the remains of the American transit expedition at Campbell Town, Tasmania, and to Dr Barry Clark, of the Astronomical Society of Victoria, who carefully checked the section on the 1874 transit observations from Victoria, Australia. Paul Brunton, from the State Library of New South Wales, kindly provided expert information and assistance on Cook's voyage.

Dan Kiselman, of the Institute for Solar Physics of the Royal Swedish Academy of Sciences, provided high-resolution versions of images from the Swedish 1-metre Solar Telescope.

And I am grateful to many other people who kindly provided images including Melissa Hulbert of Sydney Observatory, Val and Andrew White, Barry Weaver of the St Helena Virtual Library and Archive, Jonathan Bronson, Thomas Claveirole, Ian Ridpath, Julia Maddock of the Science and Technology Council of the UK, Geoff Wyatt of Sydney Observatory, Rik Davis of the Amateur Astronomers Association of New York, Hazel McGee of the British Astronomical Association, Martin Mobberley, Sally Russell, Glenn Schneider of the University of Arizona, Xiaojin (Jerry) Zhu, Johannes Schedler of the Panther Observatory and David Cortner. As well, many institutions kindly supplied images, some rarely seen. Sally Bosken of the James Melville Gilliss Library at the US Naval Observatory kindly supplied many complimentary photographs of the American expeditions in 1874. I would especially like to single out NASA for making its images publically accessible: many wonderful images from its spacecraft have been selected to illustrate this book.

At the Powerhouse Museum, I am grateful for support from the Director, Dr Dawn Casey, the manager of Editorial and Publishing, Judith Matheson, Principal Curator, Matthew Connell, and the Manager, Sydney Observatory, Toner Stevenson. Assistance from Tracy Goulding and Melanie Cariss is appreciated, as is the huge amount of work in collating and obtaining images by Damian McDonald of Curatorial, together with Iwona Hetherington of the Photo Library.

At NewSouth Publishing, the initial discussions began with Stephen Pincock. The book became possible through the assistance of the publisher, Jane McCredie, Project Editor, Melita Rogowsky, Designer, Di Quick, and Managing Editor, Heather Cam. And very special thanks are due to the book's editor, John Mapps, who pulled the work together.

Finally and most importantly, a big thank you to my wife Madelene for her unfailing support during the long period that I was occupied with researching and writing this book.

INDEX

Note: Page numbers in *italics* refer to illustrations and photographs.